AF541283

GREEN LANDSCAPE ARCHITECTURE

GREEN LANDSCAPE ARCHITECTURE

BY

CHLOE HALL
ELLA WATTS
AIMO PARKER

KNOWLEDGE BAKERS

ISBN: 9789390013524 (HB)

Published by : **Knowledge Bakers**
Pune, India
Email: knowledgebakers2019@gmail.com

Digitally Printed at : **Replika Press Pvt. Ltd.**

PREFACE

Green landscape architecture is an innovative approach to landscape design that seeks to create sustainable and environmentally friendly outdoor spaces. The philosophy of this design style is centered on promoting the use of natural elements to create beautiful, functional, and sustainable landscapes. This approach involves the careful consideration of various environmental factors such as climate, soil, water, and vegetation in the creation of outdoor spaces that are harmonious with nature. One of the key principles of green landscape architecture is the use of native plants and materials. Native plants are those that are naturally found in a particular region, and are adapted to the local climate and soil conditions. The use of native plants in landscape design helps to support local ecosystems, promote biodiversity, and reduce the need for artificial fertilizers and pesticides. This approach not only enhances the natural beauty of the landscape but also creates a more sustainable and self-sufficient environment. Another principle of green landscape architecture is the efficient use of resources. This involves using environmentally sustainable materials and design techniques to minimize the environmental impact of landscape design. For example, the use of permeable paving materials allows rainwater to seep into the soil instead of being channeled into storm water systems. This not only reduces the risk of flooding but also helps to replenish groundwater supplies. Additionally, the use of renewable energy sources such as solar panels and wind turbines can help to power outdoor lighting and other landscape features. In addition to promoting sustainability, green landscape architecture can also have significant health and wellness benefits. The inclusion of natural elements such as trees, water features, and green spaces has been shown to reduce stress levels and improve mental health. Furthermore, outdoor spaces that are designed with sustainability in mind can promote physical activity and social interaction, both of which are essential components of a healthy lifestyle.

Green landscape architecture also has economic benefits. By creating sustainable and beautiful outdoor spaces, property values can be increased, and maintenance costs can be reduced. Additionally, the use of sustainable materials and design techniques can help to reduce long-term maintenance costs and energy expenses. Green landscape architecture is a design philosophy that seeks to create sustainable and environmentally friendly outdoor spaces. The use of native plants and materials, efficient resource use, and the promotion of health and wellness are some of the key principles of this approach. By promoting sustainability, health, and economic benefits, green landscape architecture can help to create beautiful and functional outdoor spaces that are harmonious with nature.

The book covers a wide range of topics related to landscape architecture and its role in promoting ecological sustainability, community attachment, and environmental stewardship. It explores the concept of therapeutic landscapes, which are designed to promote health and well-being through a natural weaving of culture, health, and land. It presents a vision for restoring the degraded landscape of the Centre Region of Portugal, highlighting the role of sustainable land use planning in promoting ecological sustainability. Provides descriptive insights into the discipline and profession of landscape architecture, focusing on the American landscape architecture research universe and the importance of higher education in the field. It a valuable resource for students, professionals, and researchers in the green landscape architecture.

We are grateful to all those persons as well as various books, manuals, periodicals, magazines, journals etc. that helped in the preparation of this book. In spite of the best efforts, it is possible that some errors may have occurred into the compilation and editing of the book. Further queries, constructive suggestions and criticisms for the improvement of the book are always welcomed and shall be thankfully acknowledged.

Chloe Hall
Ella Watts
Aimo Parker

CONTENTS

Chapter 1

Introductory Chapter: Understanding the Ataturk Forest Farm from an Ecological Perspective

1. Introduction

The farm that has gained a significant place in the minds of the Turkish people and has been most well-known to them since the proclamation of the republic is undoubtedly Ataturk Forest Farm. Mustafa Kemal Ataturk laid the foundations of the modern Ataturk Forest Farm, initially called Gazi Forest Farm, in 1925, 5 years after the opening of the Grand National Assembly of Turkey and 2 years after the proclamation of the republic.

Following the years when Turkish people achieved epic triumphs, Ataturk Forest Farm was built during a challenging period of poverty. The Anatolian people were the main population of the Ottoman Empire and the basis of the Republic of Turkey, where 13,648,000 people [1] (76% of the population) lived in rural areas. Without a doubt, Mustafa Kemal Ataturk made the statement "Agriculture is the Basis of a National Economy" to impose the idea that the development of the Republic of Turkey should start in rural areas.

The mission of combining forestry and agriculture, two different applied sciences, was undertaken through the Ataturk Forest Farm. This mission was called the Ataturk Forest Farm because it was not solely based on farming activities and because forestry activities were more dominant than agricultural activities. Forestry and agriculture are similar disciplines but their purposes and methods are different, resembling General Surgery and Psychiatry, which both basically focus on the human body but use totally different methods and concepts. Furthermore, agricultural plans and activities are considered and conducted annually or seasonally, but forestry activities may cover a period of 100 years. Considering these differing perspectives and conditions of time, Ankara and Ataturk Forest Farm should be assessed collectively.

2. The evolution of Ankara and Ataturk Forest Farm, based on the development plan

Ankara was founded in an area that could fit into a single photo frame when it was declared the capital (**Figure 1**) [2].

Ankara is an arid city in the middle of Anatolia, the symbol of the modern Turkish Republic and its window to the world. Ankara's development plan and practices shaped the new form of the Turkish Republic, as well as the city itself.

Figure 1.
An old view from the Atatürk forest farm.

The development plan considered the impact on the city of biotic and abiotic factors, such as climate, fauna, flora, geology, geomorphology, and soil. Its approach also observed the impact of the city on people and the environment.

One of the most important factors here is undoubtedly the population projection, indicating how many people are estimated to be present in a certain place as well as the dimensions of this place. The number of people living in the Turkish Republic was 13,648,000 during those years, and the first census was performed for the first time after the proclamation of the republic was performed on October 28, 1927. Ankara's urban population was 74,784, of whom 49,439 were male and 25,345 were female [3]. Since the number of people living in rural areas of the Turkish Republic was 76% of the total population, a total population of 250,000–300,000 people was estimated for the whole Ankara area.

Carl Lörcher, a German architect and planner, had prepared two separate plans for the old and new city in 1924 and 1925 for Ankara's development between 1925 and 1930. However, the extent of his spatial planning studies and the housing conditions suggested in the plans were not adequate for forming a living area for 250–300,000 people, so a new development plan was needed. Exceeding the afore-noted population projection considered for Ankara's development plan between 1925 and 1930 forced the city to undergo spatial expansion, and new development plans were needed. Accordingly, an international contest was held and won in 1928 by Hermann Jansen in 1928 [4].

During these years, the farms called Orman, Balgat, Yağmur Baba, Macun, Güvercinlik, Ahimesud (Etimesud) and Çakırlar, constitute 80,000 decares of Ataturk Forest Farm land (102,000 decares in total), were purchased from the relevant owners using Ataturk's account number 2 in İş Bank [5]. By 1932, Hermann Jansen delivered the final development plans, and the surface area of Ataturk Forest Farm covered a larger area than the macro form of Ankara [6]. Ataturk Forest Farm reached its maximum size of 102,000 decares in the 1930s, but Ankara continued its urban development. This process can be described with the following comparison: Ataturk Forest Farm is greater than Ankara, and Ankara is greater than Ataturk Forest Farm from the period between 1930 and current times. Two important objectives of the Republic of Turkey: the agricultural development Mustafa Kemal conducted with Ataturk Forest Farm, and ensuring the urban development of Ankara that started with the plans of Hermann Jansen and that was based on modern and scientific grounds.

2.1 Agriculture and Ataturk Forest Farm

Agricultural development was one of the most important objectives of the Turkish Republic during the 1920s because most of the country's total imports consisted of main food elements, such as wheat. During those years, more than half of the total exports consisted of agricultural products, such as cotton, figs, grapes, hazelnuts, and tobacco [5].

Ataturk explained his purpose in founding Ataturk Forest Farm as follows: "The basis of the national economy is agriculture. Therefore, we attribute great importance to agricultural development. The planned and practical activities that will spread into the villages will facilitate adopting this objective. However, to gain success in this purpose, a research-based agricultural policy should be formed and an agricultural regime that every villager and citizen can easily understand and practice should be established" [7]. Ataturk reveals the notion behind the establishment of Ataturk Forest Farm with this statement.

Ataturk formed a group of Turkish agricultural experts and told them that *he wanted to establish a large farm in Ankara*, asking them to find a suitable location. One of the experts in the group stated: "We did not need to search through the entire city of Ankara and look for different natural characteristics. The reason was simple. A medieval city in the middle of an arid steppe... No trees, water, nothing... While looking for a place to establish a farm around Ankara, we focused on the current location of the farm the least. This place was not treated kindly by nature; it looked neglected and pale and made you feel pessimistic when you looked at it. The reeds in the swamp of the land through which trains pass poisoned the urban life made the people living around appear as pale and sick and acting as a source of malaria. Eagles and vultures made nests around the land, which had nothing but an adobe house back then. There was only a thin railroad as the sign of civilization and humanity on this land. When our assessments were complete, we submitted the result to the Great Chief. Ataturk showed the current location of the farm and asked: Did you visit there? We extended our common thought that the location he pointed out had none of the qualities needed for a farm and that it was only a swamp with arid sections. He replied as follows: This is what we want. A swamp, arid, poor location around Ankara. It is up to us, rather than anybody else, to physically improve that location" [5].

Ataturk's own words show that he targeted agricultural amendment before agricultural activities. Agricultural undertakings are accomplished by finding a suitable location for the pre-determined agricultural product and conducting agricultural activities in that location. However, as noted in the expert's report on Ataturk Forest Farm, Ataturk wanted to correct the land for agriculture first and only later to conduct agricultural activities.

The assessments of agriculture experts were based on the probability of performing agricultural activities on the farm. After determining the location of the modern farm land, some experts claimed it was unsuitable for any agricultural activity, but some admitted that, with stringent efforts, the land could be made suitable. For instance, Schmid, an expert from the Ministry of Agriculture, stated: "This is such a plan that either patience or money will be consumed on these arid lands and in these climate conditions" [5]. His statement is supported by the detailed soil analyses provided below, which were performed in 1926. These soil analyses indicate that the activities performed at Ataturk Forest Farm were conducted in the light of the most up-to-date data at the time.

ARID SOIL (Source [8]:)
Chemistry Laboratory
Number 32/42

1926 ANALYSIS REPORT
Sent by: Directorate of Forest Farm.
Analyzed: Soil sample.
Amount: Approximately 10 kilograms.
Result of mechanical analysis:

Main	% in the soil from which water vapor is dissolved
<0.002 millimeters	38.0%
0.002–0.02 millimeters	33.5%
0.02–0.25 millimeters	27.5%
0.25–2.0 millimeters	1.0%

Source [8].

Chemical analysis:	
Amount of carbon calcium	4.1%
Amount of fluid solution:	
Cations: Alkali ions (Na and K)	2.5138%
Calcium ion (Ca)	0.0337%
Magnesium ion (Mg)	0.0709%
Anions: Hydrocarbon ion (HCO4)	0.0419%
Chlorine ion (Cl)	0.2190%
Sulfate (So4)	2.2698%
Nitrate (No3)	0.0845%
The main carbon amount was calculated through Na_2Co_3	0.0056%
Concentrated hydrogen ion (Ph)	8.7%

Source [8].

Different experts analyzed the lands of Ataturk Forest Farm at different times, and the analyses revealed that the soil structure of the farmland was not homogeneous. Therefore, different agricultural activities were performed on the different soil types.

Agriculture is affected by all climate parameters but the amount and timing of rainfall is the most important parameter. Water should be supplied, or water resources should be accessed to supplement rainfall and sustain agricultural activities.

Accordingly, Atatürk Forest Farm interventions to access water included the following:

1. Building a barrage to irrigate the bottomlands with water from Ince Su and Bend Lake, and opening a 10-kilometer canal;

2. Building a large barrage to irrigate the other large section of the plain from Çubuk River, and opening a second canal, nine kilometers in length;

3. Building a barrage to irrigate the Tahar plain from Macun and the Çubuk River, and opening another canal, 1 km in length;

4. Building a concrete barrage to collect the surface water flowing through the Tahar strait in winter, and to irrigate the lands by collecting the underground water;

5. Collecting the underground water in Çorak River and building a pond;

6. Building an artificial pond on Kelek meadow to collect rainwater and underground water;

7. Building a dam in Istanbul that is 146 meters long, and opening irrigation canals;

8. Collecting the underground water in Çakırlar Farm through galleries, and irrigating crops in time; and

9. Purchasing five large transportable centrifuge pumps and irrigating crops throughout the farm [5].

Using irrigation increased productivity on Ataturk Forest Farm, and agricultural activities that suited the climate and land conditions were performed in certain parts of the Farm, although the infrastructure activities mentioned before were too expensive [5]. Ankara has been a model for agricultural activities. Due to the nature of agriculture, fruit productivity at Ataturk Forest Farm falls to minimum levels and significantly varies, year by year, due to climate issues, even though conservation actions are taken" [5].

3. Forestry and Ataturk Forest Farm

Ataturk Forest Farm did not focus solely on agricultural activities; it also played a key role in growing trees that were conserved under the conditions of that era. By building a forest, *a unity of life collectively formed* by trees, shrubs, bushes, and herbaceous plants of a certain height, characteristics and density, kelp, Boston ferns and fungi, microorganisms living under and over the ground, and various bugs and animals, all of which can create a specific climate in the borders of the forest, is created [9].

The afforestation of Ataturk Forest Farm was greatly assisted by the *Forestry School* [10] opened in İstanbul in 1857. The first afforestation activity conducted in Turkey dates back to 1892. Aleppo pine, cedar, black pine, gallnut, and ash trees were planted in a 20–25 decare area. The fatigue arising from World War I and after the Turkish war of independence, the rapid occurrence of reforms, and limited financial resources, all meant little importance was attributed to forestry and afforestation till 1937. Afforestation of Ankara's Ataturk Forest Farm was renewed after World War II when the Yalova-Termal afforestation and Tarsus-Karabucak eucalyptus forest were among the important afforestation activities [11]. Various laws enacted between 1923 and 1937 enabled performing afforestation at Ataturk Forest Farm, but no legal actions regarding the control of flooding were taken during those years [12]. The micro-climatic impact expected from the afforestation activities performed at Ataturk Forest Farm has finally been realized today, although conditions have changed since those early days. The necessity of a challenging study can still be mentioned if the soil analyses provided above are to be assessed by modern forestry and agriculture experts.

3.1 Ataturk Forest Farm being gifted to the treasury

Mustafa Kemal Ataturk and his farmworkers made great efforts for 13 challenging years. At the end of this period, Ataturk thought about granting his properties and goods to the Republican People's Party, which he founded, and he mentioned this idea during the congress of 1927 [5]. In 1937, Ataturk submitted a letter to the Speaker's Office at the Grand National Assembly of Turkey indicating his wish to donate the Ataturk Forest Farm, and he stated [5]: "As you know, I founded numerous farms in different parts of the country at different times to gain experiences in terms of agriculture. I give all these farms which hosted various agricultural arts as well as all sorts of agricultural products produced under relevant climates *during the challenging efforts lasting 13 years along with all equipment, animals, and fixtures to the treasury*. A brief list indicating the lands of farms, equipment, and fixtures is attached" [7].

Ataturk mentioned the 13 years of challenging effort spent building up the farm in the letter in which he donated Ataturk Forest Farm to the treasury—and the challenge and the effort remain even under modern conditions. Humanity's interest in (and opposition to, in certain cases) soil and climate continue to be manifested by forestry and agricultural activities. Nevertheless, it was the dedicated efforts of Ataturk and his farmworkers that turned this challenging process into a success story.

After 1937, the farmlands were managed by the government. In 1950, when Şemsettin Günaltay was the Prime Minister of Turkey, "The Law on the Foundation of Ataturk Forest Farm" was enacted. The 10th Article of the law states:

> *The transfer and handover of Ataturk Forest Farm as well as the real estates within the border of the farm during the time this law was enacted to natural and legal persons and expropriation of the aforenoted are subject to receiving permission through a special law. Prior to the enactment of this law, the aforenoted shall not be applied to the real estates that were sold to the official offices and organizations through the approval of the Administrative Board of State Agricultural Business Department and Ministry of Agriculture [13].*

Accordingly, some sections of Ataturk Forest Farm were sold, and thus the surface area of the farmland shrank. The farm has continued shrinking physically because land from Ataturk Forest Farm was generally transferred in four ways: through a special law, through a protocol, through renting, and without law or protocol. The percentage distribution of the transferred land is as follows [14]:

Distribution of transferred lands	**(%)**
National Ministry of Defense	65%
Various public institutions and organizations and universities	22%
Metropolitan Municipality of Ankara	8%
Worker houses, farmers, different building societies	5%

Parts of Ataturk Forest Farm continued to be transferred for approximately 100 years. This was because a population of 250–300,000 people was estimated when planning the development of Ankara in the 1930s, but the actual figures were much higher. The most unexpected result of this process is that Ankara now has a population of more than 5.5 million people. This unexpected population increase put pressure not only on Ataturk Forest Farm but also on other natural resources. What would Hermann Jansen, the person who created the development plan of Ankara in the 1930s, say when he was told that the population of Berlin where he resided would remain at 3.5 million even after a century but that Ankara, the arid city in the middle of Anatolia, would have 5.5 million people 100 years later?

Considering the conditions of the era, neither Jansen nor anybody else would believe such a claim. While this unexpected population growth expanded Ankara, it shrank Ataturk Forest Farm. The process started during the era of Ataturk's contemporaries after Ataturk passed on. This transformation was not directed against Ataturk Forest Farm; instead, it was conducted in conjunction with efforts to maintain and improve the Republic of Turkey.

3.2 Ecological impacts of Ataturk Forest Farm

The transformation of urban spaces undoubtedly causes changes in environmental processes. In a 2009 study by Bilgili, the temperatures at different areal and green locations of Ankara were measured by fixed and mobile climate stations. Ataturk Forest Farm was statistically shown to be cooler than the built-up areas around Altınpark, Gençlik Parkı and Kurtuluş Parkı and was also cooler than the other green areas. The scientific findings varied according to the green areas, the micro-scale climate change caused by tall trees with large petal leaves in and around Ataturk Forest Farm, and the spatial (size), structural (pattern and plant cover), and temporal (phenological periods) traits of green areas. Therefore, the contribution of green areas to an urban ecosystem is a dynamic process caused by the collective impact of these areas' spatial, structural and temporal traits, and it changes over time. According to the Normalized Difference Vegetation Index, the plant cover of Ataturk Forest Farm, and of the forests surrounding the agricultural areas there, have a linear form, distinctively separate from the agricultural areas [15].

The trees planted approximately a century ago still have an impact on the climate parameters of Ankara in current times. Because it is in the center of the city, Ataturk Forest Farm's spatial traits that will create Ankara's green area system for the next century should be reassessed at this point.

References

[1] TURKSTAT. General Censuses. Turkish Statistical Institute; 2020. Retrieved from: http://www.tuik.gov.tr/PreTablo.do?alt_id=1047 [Accessed: 26 May 2020]

[2] Memlük Y, Keskinok Ç, Dinçer G, Kılıç C, Şenoğlu K, Erkan K. Bir Çağdaşlaşma Öyküsü Cumhuriyet Devriminin Büyük Eseri Atatürk Orman Çiftliği. Koleksiyoncular Derneği: Ankara; 2007

[3] Toprak Z. Cumhuriyet Ankara'sında İlk Nüfus Sayımı (Tecrübe Tahriri - 1927). Ankara Derg. 1991;**1**(2):57-66

[4] Burat S. Yeşilyollarda Hareketle İstirahat': Jansen Planlarında Başkentin Kentsel Yeşil Alan Tasarımları ve Bunların Uygulanma ve Değiştirilme Süreci (1932-1960). İdealkent Cilt. 2011;**4**:101-102

[5] Öztoprak İ. Atatürk Orman Çiftliği'nin Tarihi. Ankara: Atatürk Araştırma Merkezi. 2006;**14**:25-35

[6] Kimyon D, Serter G. Atatürk Orman Çiftliği'nin ve Ankara'nın Değişimi Dönüşümü. Cilt. 2015;**25**(1):51

[7] Anonim. Atatürk ve Tarım. Antalya: Akdeniz Üniversitesi Ziraat Fakültesi; 2020

[8] Anonim. ODTÜ Mimarlık Fakültesi AOÇ Araştırmaları. Ankara: ODTÜ; 2020. Retrieved from: http://aocarastirmalari.arch.metu.edu.tr/belge-yapitlar/.s.12

[9] Anonim. Türkiye Orman Varlığı. Ankara: Orman Ve Su İşleri Bakanliği Orman Genel Müdürlüğü; 2014

[10] Anonim. İstanbul Üniversitesi Orman Fakültesi Tarihçe. İstanbul: İstanbul Üniversitesi Orman Fakültesi; 2020

[11] Arslantaş M. Atatürk Orman Çiftliğinde Kızılçam (Pinus Brutia), Karaçam (Pinus Nigra), Sedir (Cedrus Libani) Türleriyle Yapılan Ağaçlandırma Çalışmalarının 6 Yıllık Sonuçlarının İrdelenmesi. T.C Isparta Uygulamalı Bilimler Üniversitesi. 2019;**1**

[12] Mustafa Dal, "Türkiye'de Yandere Islahı Çalışmalarında Örgütsel Yapı", İstanbul Üniversitesi Orman Mühendisliği Anabilim Dalı Orman İnşaatı ve Transportu Programı Yüksek Lisan Tezi. 2011. s. Viii

[13] Türkiye Cumhuriyeti Resmi Gazetesi. 5659 sayılı Atatürk Orman Çiftliği Kurulması Kanunu. Ankara; 1950

[14] ATAK E, ŞAHİN SZ. Atatürk Orman Çiftliği'nin 79 Yılı ve Çiftliğin Korunmasına Yönelik Politika Arayışları. Planlama. 2004;**3**:80-88

[15] Bilgili BC. Ankara Kenti Yeşil Alanlarının Kent Ekosistemine Olan Etkilerinin Bazı Ekolojik Göstergeler Çerçevesinde Değerlendirilmesi Üzerine Bir Araştırma. Ankara Üniversitesi Fen Bilimleri Enstitüsü Doktora Tezi; 2009. pp. S.1-165

Chapter 2

Erosion Quantification and Management: Southeastern Nigeria Case Study

Abstract

Soil erosion in Southeastern Nigeria is assuming an unusual dimension despite efforts by successive governments to control the phenomenon. Agronomic activities on eroding surfaces can give rise to landscapes much different from the original. Research activities in erosion quantification, the findings and how their applications have contributed to soil erosion management are highlighted. A key factor is the community efforts which have been relegated to a top-down approach occasioned by land use, land tenure and technological changes. The system is often a preventive management approach which achieves ecological and economic benefits. This chapter also discusses the indigenous methods of soil conservation and proposes their inclusions for sustainable management. To manage soil erosion in the region, emphasis must be placed on preventive management rather than crisis-management. Such approach will ensure that fewer resources are expended and land is appropriately conserved. To this end, soil can play its many environmental roles adequately.

Keywords: soil erosion, indigenous knowledge, soil conservation, erosion quantification, land use

1. Introduction

Methodologies for sustainable management of land degradation, economic growth and poverty reduction have become topical issues in present African research activities because of the danger posed by their neglect [1]. Land degradation especially the soil erosion aspect has been recognized as a serious threat to environmental sustainability. It impacts life on earth through degradation of land resources, loss of farmlands, decline in soil fertility due to top soil losses, contributes to climate change due to a compromise in soils C-sink potentials. In lowlands, eroded soils are often deposited as sediments on both land and river bodies. Thus further impoverishing rural communities who are often ill- equipped to manage the threat on land and water resources. As a result of the many implications of soil erosion on the environment, many efforts have been made to adequately understand the phenomenon so as to better manage it. Many of such efforts have failed due to little consideration of the several factors and their environmental peculiarities. The factors include: rainfall, soil properties, topography, and land-use and management. Since soil erosion begins in the farmer's field, scientific results could

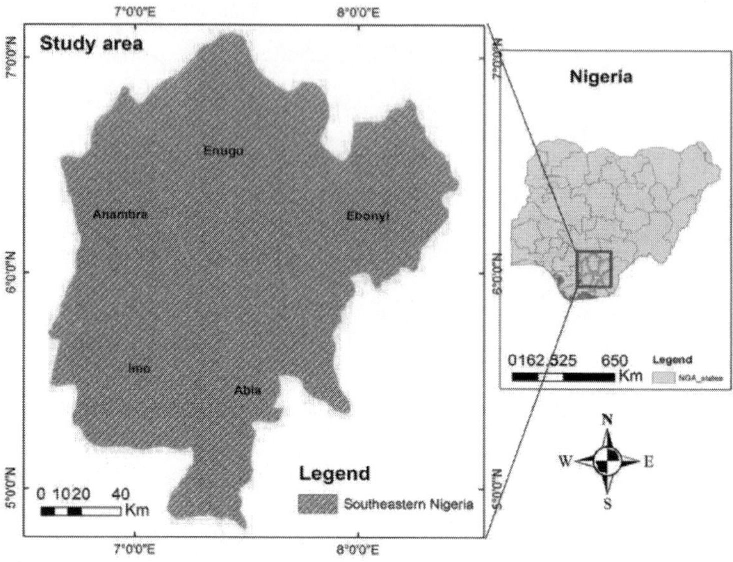

Figure 1.
Location of the reviewed area showing the states

be complemented by indigenous knowledge interventions in order to achieve better results. Local people have not only found nomenclatures for their soils, the indigenous knowledge has also extended to practices that have been through longterm observation of their interaction with the environment and transferred from generation to generation. Ezeaku and Salau [2] defined indigenous soil knowledge system as "the knowledge of soil properties and management possessed by people living in a particular environment for a long period of time. Soil conservation practices applicable to the North- Western zone of Nigeria include: contour farming, ridge tying, strip cropping, crop rotation, planted fallows, conservation pits, crop livestock farming and adequate fertilizer use [3, 4]. Howevever not much studies have been carried out in the Southeastern Nigeria (**Figure 1**) which is particularly an erosion prone zone.

2. Soil erosion and landscape evolution

Landscapes evolve under the influence of a complex suite of natural processes, many of which may be either directly or indirectly influenced by land use [5]. For example, a long history of cultivation can leave significant footprints on the original landscape. Under an unfavorable land-use condition, soil particles are moved from one position to another through agents of erosion such as wind, water and gravity. The series of particle movement consisting of detachment, transportation and deposition contribute to the evolution of landscapes. This is more so when the phenomenon occurs at accelerated dimensions as a result of continued anthropogenic activities. The effect of the denudation is a drop in soil surface level [6]. Future landscape evolution driven by soil erosion is expected to be exacerbated by land-use change, agricultural intensification and climate change [7, 8], coal mining and exploration [9], terracing [10], other mining activities, road networks and agricultural practices [11]. With current efforts to make resources available to the ever increasing population, the impact of these events in terms of magnitude and impact is expected to increase and destabilize geomorphic systems. In Southeastern Nigeria, Nwajide [12] observed that a major exogenic geological hazard is soil loss due to sheet and gully erosion. Of which the sheet type occurs as more or less removal of topsoil by flood but does not appear to threaten agricultural production or human habitation. Nevertheless, during major floods, sheet erosion may threaten small holders' food production for a period of time (**Figure 2**). However, its impact is often considered to be obliterated by the rapid rate of soil regeneration. The gully types have been observed to be the most obvious because of the remarkable impression they leave on the surface of the earth [13]. A gully is a distinct channel carved by running water into an unconsolidated substratum, and through which water flows only during and immediately after heavy rains. They are also a visible manifestation of the physical loss of the land due to erosion (**Figure 3**). Idike [13] observed that most studies of soil erosion in Southeastern Nigeria had strong focus on gully incision and gully prone regions but less on the little noticed sheet erosion. Such erosion contributes to land degradation but often in slower dimensions that land users fail to notice and are they often occur alongside with gullies in erosion prone areas. It is quite clear that soil erosion alters hydrology and landscape and connectivity patterns therefore necessitating efforts towards its better quantification.

Figure 2.
Inspection of lowland sites affected by sheet erosion following the 2012 flood at Odekpe, Southeastern Nigeria.

Figure 3.
A gully incised landform at St. Francis Enugwu-Ukwu, Southeastern Nigeria.

3. Soil erosion quantification techniques

In Nigeria, the detailed history of erosion quantification evolution may have been lost. However, there are remarkable timeline of events that are observable from literature. Earlier (1930–1955), soil erosion studies were descriptive, involving much field surveys and subsequent mapping. It was dominated by geographers and geologists who made efforts to understand the soil and its environment at regional scales (for example, see [14–16]). The second phase of soil erosion researchers (1955–1985) were mainly agriculturally-inclined (agronomists and soil scientist) with fewer geomorphologists, who established runoff plots, simulated soil loss using desurfacing approaches and a host of other techniques which were experimented in order to understand soil-water, erosion-productivity interactions in fields [17, 18]. The transitional period (1985–2000) focused on attempts to integrate technology towards broadening the scale of soil erosion research [19–24]. Presently (2000-date), soil erosion research has been multidisciplinary, multidimensional (local, regional and global) and with strong links to global issues such as social inclusion, sustainability and climate change. It is worthy to note that researchers continue to apply different methods depending on their objective and no method is obsolete Per se but a compromise of the other. The methods that have been applied by researchers in Southeastern erosion quantification are summarized in **Table 1**. The common erosion quantification methods are discussed below.

3.1 Mapping and direct field observations

Photographs can be used for detecting morphological change at varying scales and for recording the spatial relationship of landforms in order to provide three-dimensional information that can be used to construct Digital Terrain Models (DTMs) (for example, see [31]). It is also carries supplementary details useful for interpreting erosion rates or patterns. Maps produced as a result of detailed reconnaissance land resources surveys generally also contain information on the erosion hazard and on evidence of past erosion. Areas affected by sheet-rill-and gully- erosion can be recognized on aerial photographs and the growth of the erosion affected areas or the effects of conservation measures can be be traced from available maps or photographs and additional information collected in the field. Information obtained in the field may include using a simple scoring system to rate the severity of the erosion from e.g. the exposure of tree roots, the surface crusting, the thickness of the A horizon, erosion forms and shapes etc. Onweremadu [32] used field sampling aided by morphological landscape changes to identify erosion units for conservation treatments. However, field surveys can be time consuming but with the development of remote sensing techniques, more efficient methods of obtaining spatiotemporal erosion information are emerging. The disadvantages of using mapping as a tool for assessing soil erosion are the needs for cartographic skills, challenge of ascertaining whet difficulties in interpreting whether the current situation of the erosion phenomenon time constraints and variations in map quality.

3.2 Runoff plot studies

Runoff-plot methods are designed by using artificial boundaries to define a plot area and sediments are collected from a receptacle downslope. They could also be closed plots systems which uses rainfall simulators to study erosive events or open systems. Runoff plots are valuable research tools in soil erosion and surface runoff (soil loss) studies, evaluating conservation measures, effect of different crops and

Location	Erosion Type	Quantification method	Result	Management recommendation	Author(s)
Abia State	gully	GPS and GIS	False bedded sandstone formed the major area for gully dispersion especially where slope was greater than 15 degrees	Proper land use	[25]
Onitsha, Port Harcourt, Owerri, Enugu, Uyo, Calabar, Ikom and Ogoja	Sheet	USLE, Rainfall erosivity	Severe erosion causing rains are associated with the rainy seasons; Calabar, Owerri and Port Harcourt had very high erosivity index	Monitoring hydrologic and climate related factors as well as land-use management	[26]
Anambra	Sheet	GIS and RUSLE-based	Mean value of estimated soil loss of 214.82t ha-1.;High rainfall erosivity combined withhigh slope factor and decreasing vegetal cover	Comparison between estimated and measured soil loss	[27]
Anambra and Enugu States	Sheet, interill and rill erosion	SLESMA and USLE erosion models	Coincidental and significant relationship between the USLE estimated maps and extent of actual gullying on ground	assesment of individual soil erodibility and not rating based on soil taxonomy	[24]
Orlu-Okigwe Asix of Imo State	Gully	Field and Landsat data	Surface phenomenon of washing away of loose top soils is not the only factors responsible for gullies but also deep-seated near-surface structural weakness along gully axis.	Emergency environmental monitoring	[28]
Anambra Basin	Gully	Field, Remote sensing and Geotechnical analysis	Development of gullies on steep slopes and non-vegetated areas are facilitated by cohensionless and very permeable nature of Ajali and Nanka sandy formations	Agronomic and engineering practices could help mitigate formation and expansion of gullies	[29]
Orashi Catchment, Anambra State	Not known	Field, expert judgment and remote sensed data	Vegetation and slope are the main factors governing erosion susceptibility	Change in landuse to a more integrated basin development system and public awareness on soil conservation and strategic planning	[30]

Table 1.
Summary of some erosion quantification approaches and management recommendations in Southeastern Nigeria.

management practices. They are commonly used to monitor hillside erosion but the design of runoff plots (in terms of plot dimension, runoff and erosion collection system, methods to monitor sediment concentration etc) are not standardized, making their results technique dependent [33]. Li *et al.* [34] anticipated that runoff research will tend to be more precisely location and model-inclined, technologically advanced and quantitatively precise in future. Iwara and Ewa [35] constructed erosion plots on natural fallow vegetation varying ages in southeastern Nigeria. They observed that July to September experienced highest amount of runoff and sediment losses. A better performance of the 10 and 3-year old fallow over the 5-year old fallow lead them to the conclusion that surface cover type and extent had greater influence over erosion processes than the age of fallow. The use of runoff-plot often alters the natural hydrology of fields due to their artificial boundaries and therefore may not accurately represent the actual erosion conditions. Extrapolation of the plot scale experiment beyond the area of observation may also be erroneous.

3.3 Erosion pin technique

The erosion pin method is a simple and feasible approach for soil erosion monitoring by inserting rods or nails into surface of slopes and using the basis of length of pin exposed or movement of washer placed on the pin. The technique has been successfully been modified and its photo-electronic erosion pin (PEEP) modification was efficiently used to monitor stream bank erosion by Lawler in 1989 [34, 36]. Erosion pin can aid in dynamic monitoring of the initial stage of gullying by identifying surface roughness, detachment and deposition. It can also conveniently monitor bank collapse and other short-term field monitoring. Some of its limitation include: susceptibility to environmental and human interference, need for close contact with assessed land and small range of observation. It can also be used to monitor gullies and landslides.

3.4 Erosion marker technique

Erosion markers allow carrying out analyses at larger temporal and spatial scales than those that are achieved through experimental plots. Bio-markers such as tree ring characteristics have been used to estimate the rates of soil erosion from decennial to millennium time scales by applying dendrogeomorphology [37]. The original landscape in relation to exposed roots can be a marker of soil erosion processes. However, as it is not always easy to identify the original land surface level, the vertical distance from an exposed root to the present ground surface may represent an underestimation of the total depth of the material [38]. The use of biomarkers is useful for long-term erosion quantification but it is subject to errors due to the natural variability of plants.

3.5 Radionuclide tracer method

Over the last few decades, geochemical methods have also been used to quantify erosion rates at different temporal scales. Examples of radionuclides which have been used as erosion tracers include ^{137}Cs, ^{210}Pb and ^{7}Be. The application of environmental radionuclides in soil erosion surveys is based on the premise of adsorption and redistribution of fallout by soil and sediment particles following erosion and sedimentation [39]. Radionuclide observations showing losses compared to the reference value indicate erosion. Observations greater than the reference value shows deposition. Unfortunately, this approach is yet to be applied to Southeastern Nigeria. Its first application in Ibadan, Southwestern Nigeria was reported to be a valuable alternative to conventional methods for soil erosion for obtaining quantitative data on soil erosion and deposition [40].

3.5.1 ^{137}Cs tracer method

^{137}Cs tracer technology rapidly developed and became the major means for monitoring soil erosion, determining soil erosion and sedimentation rates, quantitatively analyzing soil net loss, and other applications in the field [34]. The principal limitations of the ^{137}Cs approach include the costs of analytical equipment and the difficulties experienced in interpreting medium term estimates of average soil redistribution rates in the absence of complementary information on land use patterns and intensity. The method also requires a long measurement time and a high cost of laboratory analysis.

3.5.2 $^{210}Pb_{ex}$ tracer method

According to Li *et al.* [34], the $^{210}Pb_{ex}$ tracer method can distinguish the changes in atmospheric particles and human causes of trace elements, the reconstruction of pollution sources, and the history of river deposition and erosion in the past 100 years. However, its limitations include complex sample processing, high accuracy requirements, and difficulty in obtaining the flux of deposition for a particular year. In future, development of this technique is to further improve the quantitative relationship between the amounts of $^{210}Pb_{ex}$ loss and soil erosion. Its combination with other tracers and models is also likely.

3.5.3 ^{7}Beryllium tracer method

Li *et al.* [34] observed that this method can be applied to evaluate soil erosion under a particular intensity of land use, thereby providing an important basis for the monitoring and control of soil erosion. However, the application of ^{7}Be tracing still has some problems. For example, the shallowness of ^{7}Be distribution complicates sampling. It is, however, important to recognize that the use of ^{7}Be measurements is best suited to situations where significant erosion events are separated by 5 months in order to minimize the effect of previous erosion [39]. Since ^{7}Be could reflect the effects of soil erosion factors. Therefore, the scope of applications of the ^{7}Be tracer method can be broadened to explore the comprehensive effect on specific small watersheds based on the hydrological and meteorological conditions of soil erosion [34].

3.5.4 Magnetic tracer method

The application of magnetic tracers has two aspects: (i) to trace sediment sources using magnetic minerals in the environment, (ii) to indicate environmental change in basins. Hence Li *et al.* [34] synthesized that the magnetic tracer method can reflect the history of land use pattern, vegetation succession, and soil erosion in a watershed. It can also identify the soil distribution and the erosion rate for certain period. Therefore, this method can be used to provide a theoretical basis for soil erosion prediction and monitoring, and a history of the development of small watersheds. The advantages of this method are the transportability of the equipment, the methods simplicity; meeting the need of large samples and non-destructive nature. The method is however constrained by inability to trace magnetic properties and depth of soil erosion or deposition. Presently, magnetic tracers have been used to study soil formation, classification of soils, and the quantitative description of evolution, occurrence, and development of erosion.

3.6 Soil erosion models

Soil erosion models are quantitative approaches in study of soil erosion. Based on literature searches, the application of models in Southeastern Nigeria is still minimal. Models can be classified into three groups viz. Empirical, Physically-based and Conceptual (partly empirical/mixed) [41].

3.6.1 Empirical statistical model

Empirical models are based primarily on observation and inductive logic from the environment. Empirical models such as the Universal Soil Loss Equation (USLE), Revised Universal Soil Loss Equation (RUSLE), the Unit Stream Power Based Erosion/Deposition model (USPED), Erosion Productivity Impact Calculator (EPIC). Empirical models are mainly based on the USLE and remain widely used till date even in regions with limited data. Li *et al.* [34] documented the advantage and disadvantage of these models to include: (1) the formula is concise and the meaning of each factor is clear. (2) The calculation method of the factor has been basically mature and the parameters are easy to obtain for the continuous improvement and perfection of the model. (3) After several years of verification and testing, the accuracy of the model meets the needs of the application. In the tropics, the incompetence of the USLE in its rainfall erosivity component has been overcome by the incorporation of rainfall erosivity values in the RUSLE [27]. Applying the equation in the Anambra area, they observed that about 1804.39 km^2 (39.49%) of the area had slight erosion rate of 0–10 t/ha/yr., while rates of erosion in 746.60 km^2 (16.34%), 1025.38 km^2 (6.28%) and 45.59 km^2 (1.02%) of the area are 10.6–85.3, 85.4–235.2, 235.3–608, 608.1–2200 and > 2200.1 t/ha/yr. respectively. They noted that high rainfall erosivity, moderate to high slope and decreasing vegetal cover were the major factors driving soil loss in the area. In an earlier study, Igwe *et al.* [24] compared the USLE and the Soil Loss Estimation Model for South Africa (SLESMA) in producing soil erosion working maps in Anambra and Enugu States, South-East Nigeria. They found out that the USLE model reflected better the actual field situations except for its high values, absolute values compared to the global scale. The values were categorized into very slight (<50 Mg/ha/yr); slight (50–150 Mg/ha/yr); moderate (151–500 Mg/ha/yr); severe (501–1500 Mg/ha/yr) and very severe (>1500 Mg/ha/yr). A similar high value of above 200 t/ha/yr. was reported in Uyo metropolis, Nigeria by Fashae *et al.* [42] who observed that the values corresponded with areas with active gullies and altered vegetation cover. Obinna *et al.* [43] applied RUSLE model on the entire Southeastern Nigeria and observed that the results corresponded with known areas of gully menace in the region. Most of the erosion hotspots were located around the north-eastern part of the region covering most parts of Ebonyi State, some parts of Enugu State (Northwest axis), Anambra State (South East and Central axis), and most parts of Abia State. It could, therefore, be concluded that the high number of active gully occurrence may translate into the likelihood of other forms of erosion in the tropics.

3.6.2 Physical process model

The physical process model is based on the study of the processes and mechanisms of soil erosion e.g. stream flow or sediment transport. Examples of physical models include Water Erosion Prediction Project (WEPP), European Soil Erosion Model (EUROSEM). The WEPP model can simulate soil erosion, non-regular steep slope, and soil, tillage, and management measures by calculating the temporal and spatial distribution of soil erosion and predicting the movement of sediment in the slope and basin [34]. The WEPP model reflects the applicability and ductility of the temporal and spatial distribution of erosion and sediment; thus, numerous scholars still use this method. Although the physical model greatly compensates for the defects of the empirical model, this approach also has some shortcomings. (1) The physical mechanism of soil erosion is relatively complex and unclear. Some parameters in the physical process model are still dependent on the empirical model. (2) The large range of the study area is the major obstacle that hinders the use of the model because of the exacting demand of the model parameters. (3) The structure of the physical process model is complex and may change because the form has not been unified.

3.6.3 Conceptual model

Conceptual models lie somewhere between physically-based models and empirical models, and are based on spatially lumped forms of water and sediment continuity equations. Parameter values for conceptual models have typically been obtained through calibration against observed data, such as stream discharge and concentration measurements [44]. ANSWERS (Areal Nonpoint Source Watershed Environment Response Simulation), CREAMS (Chemicals, Runoff and Erosion from Agricultural Management Systems), and MODANSW (MODifiedANSWers) are basically conceptual and event based models [41].

3.7 Remote sensing

Remote sensing allows detection of erosion in large areas using aerial photographs and satellite remote-sensing data without disturbing the studied land area. The assumptions in the use of this method include, first, that the geomorphic processes of interest produce detectable changes in the spatial or temporal pattern of electromagnetic radiation and, secondly, that any geometric distortions arising from the sensor can be discriminated from real changes in landscape features [39]. Quickbird, SPOT 5 and IKONOS are very promising for identifying erosion features, such as individual gullies. NigeriaSat-1 image data and Landsat ETM data have been applied for a comparative classification of landuse patterns and gully development in southeastern Nigeria [45]. There are also opportunities in obtaining data using cheaper sources such as unmanned aerial vehicles (UAVs), ordinary cameras, and smartphones. Due to the capability of this method to provide spatial and often real-time data over large areas, it is adjudged the best method suitable for Southeastern Nigeria. It is hoped that its potential will be more explored in future.

4. Factors of soil erosion

Ofomata [19] viewed the factors of soil erosion as two major components: physical (geological or "natural") and anthropogenic (human or "accelerated"). Highlighting that the human component is often exaggerated and the physical component underestimated, he divided the physical factors of soil erosion into four: climate (mainly rainfall), surface configuration (relief/slope), surface materials and vegetation. Igwe *et al.* [24] recognized rainfall, topography/relief, soil factors (geology and soil characteristics), vegetation, land use and management as the main agents that determine the extent of soil erosion hazard. The factors affecting soil erosion by water is commonly expressed in the Universal Soil Loss Equation (USLE) (Eq. (1)) as a multiplicative equation counting six environmental factors:

$$A = R \times K \times L \times S \times C \times P \tag{1}$$

Where A is the mean annual soil loss (metric tons per hectare per year), R is the rainfall and factor or rainfall erosivity factor (mega joule millimeters per hectare per hour per year), K is the soil erodibility factor (metric tons hours per mega joules per millimeter), L is the slope length factor (unitless), S is the slope steepness factor (unitless), C is the cover and management factor (unitless), and P is the support practice factor (unitless).

4.1 Rainfall erosivity factor R

The R-factor is the sum of individual storm *EI*-values for a year averaged over long time periods (>20 years) to accommodate apparent cyclical rainfall patterns [46]. The *EI* term is an abbreviation for energy multiplied by the maximum intensity in 30 minutes. In the humid tropical environments, rainfall amounts and intensities often exceed the infiltration rate of excessive runoff. Ojo-Atere *et al.* [47] opined that the phenomenon was common in cultivated fields where at the peak of the rainy season, intensities of rainfall often exceed the infiltration rate of 25 mm/hr. and the soils are also nearly saturated throughout the rainy season. In fact, an earlier study by Roose [48] attributed the severe erosion damage of bare soils in the tropics to the special erosivity of the tropical rainfall rather than the ferrallitic or ferruginous soils. Salako [49] evaluated the temporal variation of rainfall erosivity between sub-humid zone (Ibadan) and the humid zone (Port Harcourt) of Nigeria. He observed a strong positive relationship between rainfall erosivity and rainfall amount.

According to Salako [50], data required are such that rates of rainfall at short-intervals (preferably ≤15 minutes) must be known, and these are very rare in many developing nations. Note that although EI_{30} is recommended by RUSLE, E_{15} was recommended for the tropics to avoid underestimation of the R-factor. The trends in rainfall erosivity have been generally evaluated using commonly available annual rainfall amount data. Lal [17] postulated a combination of daily rainfall (A) amount and maximum intensity (I_m), expressed as AI_m as a reliable index for evaluating index of tropical rainfall. Obi and Ngwu [51] observed that Lal's index of AI_m had an advantage over other indices of KE > 1 and EI_{30} in Southeastern Nigeria. Extensive studies by Igwe *et al.* [24] applied a method proposed by Arnoldus [52] to calculate the R-factor of USLE because autographic rainguage was not present in the study location and this gave the equation an advantage over the other equations. This method used monthly rainfall data to construct sub-annual R factors and then aggregated the R factors to an annual scale (Eq. (2)). It was modified from Fournier [53]'s map of the theoretical risk of erosion in Africa based on the damaging effect of precipitation.

$$FI = \frac{pi^2}{P} \tag{2}$$

pi is the average precipitation in the wettest month of the year, *P* is the mean annual total of rainfall.

Due to unsatisfactory results in West Africa, the Fournier index was modified. The modified index is given as Eq. (3):

$$MFI = \sum_{i=1}^{12} \frac{pi^2}{P} \tag{3}$$

Where MFI is the modified Fournier index, *pi* is total monthly rainfall and *P* is the total annual rainfall.

In West Africa, Eq. (4) is best to determine rainfall erosivity.

$$R = 5.44MFI - 416 \tag{4}$$

where R is the rainfall erosivity factor (mega joule millimeters per hectare per hour per year) and *MFI* is the modified Fournier index.

4.2 Soil erodibility factor K

This factor relates to the rate at which different soils erode, due to inherent properties Generally, soil properties which affect detachability include; particle size distribution, organic matter content, soil moisture, presence of cementing material such as Fe and Al oxides, stability of aggregates, clay mineralogy, rock fragments and balance of cations on the exchange complex, permeability, soil structure and strength [54]. In southeastern Nigeria, clay content, level of soil organic matter (SOM) and sesquioxides such as Al and Fe oxides, clay dispersion ratio (CDR), mean-weight diameter (MWD) and geometric-mean weight diameter (GMD) of soil aggregates were observed to influence soil erosion hazards [55]. Different parent materials were studied by Obi *et al.* [56] using four (4) methods: wet-sieving method, the Wischmeier nomograph, portable rainfall stimulator and runoff plot measurements. They recommended that the nomograph approach were unsuitable for soil erodibility studies in Southeastern Nigeria. The influence of geology on soil erodibility has been noted. For example [55] reported that sites with the worst catastrophic gullies in the classical gully sites the whole of sub-Saharan Africa exists in Southeastern Nigeria on sandy geological formations of False-bedded sandstone, Coastal Plain sands, Nanka Sands and the Bende Ameki compare to their Shale formation counterparts. Nwajide [12] observed that most soils in Nigeria bear the property of the underlying parent material from which they were formed. This follows the behavior of the soil under erosive conditions. For example soils formed on limestone, dolomite and igneous rock were more resistant than soils of sandstone and clay sedimentary formations [24]. However, information on erosion categories of various sedimentary formations of South East Nigeria is rather scanty.

According to [47], soils in the tropics with high sand contents (>60%) and low silt and clay values (<12%) and (<40%) respectively are highly erodible. Also, the weak, fine crumb surface horizon and weak subangular subsurface horizons of the former increases its vulnerability to erosion. In contrast, [24] noted that both large and fine particles were more resistant to transport because greater forces were required to entrain the former and the resistance due to cohesiveness of the latter.

4.3 Topograpy factor LS

LS reflects the influence of length and steepness of slope on soil erosion, it determines the behavior of the surface runoff. It is defined as the distance from the point where overland flows starts to the point where either the slope steepness decreases to such an extent that deposition occurs, or where surface runoff enters a well-defined channel. According to Ojo-Atere *et al.* [47], topography modifies soil profile development in three ways: (1) by influencing the quantity of precipitation absorbed and retained in the soil, thus affecting soil moisture relations, (2) by influencing the rate of removal by soil erosion and (3) by directing the movement of materials in suspension or solution from one area to another. The thinness of the solum, less organic matter and less distinct horizons than soils on level or undulating topography has been attributed to erosive exposure of the lower horizons due to slope steepness. In Southeastern Nigeria, soil erosion can occur even at slope of 5% as highly friable sandstones from the upland yields to detachment due to concentrated runoff [19]. Even in highlands or cuestas with somewhat stable lithology and erosion resistance, aggressive runoff from them devastates the lowland areas especially at the toe slopes and river head-waters [57].

LS is expressed as a unitless ratio with soil loss from the area in question in the numerator, and that from a standard plot (9% slope gradient, 22.13 m slope length) in denominator. Although L and S factors can be determined separately, the problem has been simplified by causing the L and S factor and considering the two as a single topographic factor [58]. Eq. (5) below considers the effect of L and S factors:

$$L = (\lambda / 22.13)^{m} [65.4 \sin^2 Q + 4.56 \sin Q + 0.065] \quad (5)$$

where λ is the slope length (in meters) and m is an exponent factor equivalent to 0.5 for slopes steeper than 5%, 0.4 for slopes between 3–4%, 0.3 for slopes between 1–3% and 0.2 for slopes less than 1% (based on a Wischmeier' nomograph) and Q is the slope angle.

4.4 Crop management factor C

Erosion and runoff are markedly affected by different types of vegetative cover and cropping system (vegetation type). The factor is defined as the ratio of soil loss from a field with a particular cropping and management to that of a field with a bare, tilled soil. The factor ranges from 0 to 1.0, a value of 0 indicating a 100% protection of the soil against erosion and 1.0 where there is little soil cover (e.g. freshly graded bare soil on construction site) [59]. Vegetation intercepts raindrops by facilitating infiltration of water, improving organic matter soil composition, thereby ensuring minimal erosion. The stage of growth of the crop will influence the management need (e.g. fertilizer), ability to hold soil together and canopy protection. Landuse activities that deprive soil surface of its vegetation, contributing directly to sliding, slumping, sheet and gullying include; road construction, sand mining, urbanization, industrialization and general infrastructural development [19].

4.5 Erosion control practice factor P

The erosion control practice factor P is the ratio of soil loss under a particular practice compared with the soil loss occurring under normal tillage. It therefore accounts for the positive impacts the support practice. Control practices reduces erosion potential by influencing drainage patterns, runoff concentration, runoff concentration, runoff velocity and hydraulic forces exerted by runoff on soil [60]. This factor ranges from 0 to 1 and is 1 where there are no support practices an 0 under good conservation practice. The conservation measures usually included in this factor are contouring, contour strip cropping, grassed waterways, terracing and surface mulching. Conservation measures like conservation tillage, crop rotations, residue management etc. are incorporated in the C-factor [59]. The effectiveness of conservation practices and thus the value of the P-factor generally depends on the slope steepness.

5. Landscape management in focus

In colonial times, the British Government worked on natural resource management as interest was high in expanding commercial farming enterprises. The practices were often implemented without consideration of the natural schemes used by local people to protect their soils from erosion and fertility declines. Trees were exploited without ecological considerations and conservation approach was a top-down type. Most farmers quickly abandoned such conservation model. However with increasing resource demands, most farmers are presently using unsustainable farm practices. As such, soil erosion occurrence is exacerbated. However, little study has attempted to understand occurrence of erosion in farmer's field yet sheet erosion, rills and gullies occur there. Local people with experience have recognized the peculiarity of their soils and adopted practices that are suited to their soils. The practices identified include; ditches, water harvesting 'umi', ridging, agroforestry, manure application, mulching, soil stabilizing stones, multiple cropping and embankments. Others are fallowing, conservation tillage etc.

5.1 Some soil and water conservation practices observed in Southeastern Nigeria

5.1.1 Traditional drainage ditches

These are structures constructed by digging deep within the farm or outside so as to divert runoff water and debris before reaching the farmland (**Figure 4**). It also serves as a watercourse, channeling runoff into ground water with minimal sheet erosion. This practice reduces the distance over which runoff travels over farmland and reduces water logging conditions. Morgan [61] reported that such drainage practices are constructed along the slope, often covered with grass to prevent destruction, and primarily installed in areas with high rainfall rates. The depth of this structures are variable depending on the slope of the land and need of the farmer. Locally, farmers in the study area call the traditional ditches "Umi".

5.1.2 Ridges

Ridge preparation is carried out by hoe with minimal disturbance which is a practice of conservation tilllage. Ridges are constructed across the slope so as to reduce the velocity of water movement (**Figure 5**). However, in some areas farmers

Figure 4.
Traditional drainage ditches.

Figure 5.
Broken ridges across the slope allows water movement with minimal soil loss.

tend to ignorant of implication of slope direction. The ridges are low and with large spaces between. This is to facilitate ease of water flow along the ridges. Also broken ridges are observed to be conservation measures against erosion.

5.1.3 Embankment structures

Barriers made of vegetation (dried or live), short walls (earthen or concrete) (**Figure 6**) are used to reduce the velocity of runoff water which could lead to water erosion. Embankments can function as sediment filters, aid in runoff velocity reduction, infiltration facilitator and could serve as boundaries. Farmers' in the study area view this as a very important measure of checking sheet and gully erosion.

5.1.4 Manuring/residue mulch

Farmers shred parts of trees, leave the vegetative remains after harvest and sometimes add animal wastes to the soil (**Figure 7**). Eventually, the faster decomposing part serves as nutrient to restore soil physical, chemical and biological

Figure 6.
(a) Dried vegetative barrier (b) Wall barriers

Figure 7.
Animal wastes applied to enhance soil properties.

properties while the woody part is utilized as firewood. FAO [62] reported that the practice of manuring serves conservation needs by; protecting soil surface from adverse weather conditions, increasing infiltration and reducing runoff velocity, increase organic matter supply, reduces evaporation, nutrient recycling.

5.1.5 Multiple cropping/agroforestry

Most farmland in the study area consist of woody perennials and annuals in the same land unit. Examples of annuals observed include; Mango, Citrus, Oil palm, Banana, Plantain etc. (**Figure 8**). Young [63] attributed the potential of agroforestry as an erosion control measure to its capacity to supply and maintain a good soil surface cover by the tree canopy and the pruning material. The deep roots will helpto stabilize the soil and increase moisture absorption through transpiration by the trees and crops [62].

5.1.6 Rainwater harvesting

Reservoirs-like constructions, constructed for the purpose of storing the surface run-off, generated from the catchments area. In the study area, this practice began a long time ago and is still being practiced with modifications such as larger collection reservoirs. The harvested water does not only serve irrigation purposes and

Figure 8.
Agroforestry practices common in the study area.

Figure 9.
Stones incorporated into the soil so as to stabilize it against erosion.

erosion prevention, it also saves cost for the farmer's family. Other uses include; storing nutrients and rich soil materials as well composting etc. The reservoirs used for water harvesting vary in size depending on the need of the farmer.

5.1.7 Surface roughening

Farmers incorporate stones, charcoals, kernel shells, pieces of wood etc. so as to roughen and stabilize the soil against the effect of water erosion (**Figure 9**).

6. Conclusion

Soil erosion and other anthropogenic activities can leave remarkable footprints on earths' surface. In erosion prone landscapes, the earth's surface is continually evolving due to events and processes such as (i) road and foot tracks, (ii) land use and land cover changes, (iii) hydrogeological dynamics (iv) soil erodibility etc. To understand soil erosion, efforts have been made to assess and quantify the phenomenon. However, literature on soil erosion research in Southeastern Nigeria is still minimal and some promising techniques are yet to be tested. This may be partly the reason why soil erosion related land degradation remains critically high amidst scarce and often inaccurate data. Another reason is the non-inclusion of the sustainable home-breed sustainable solutions into erosion management plans. Such knowledge could assist in both research and practice in developing robust environmental conservation approaches; that is cognizant of cultural and socioeconomic situation of the area. Local people knowledge of the soil has helped to protect soil from erosion while ensuring that its fertility is maintained. When improved, better conservation outcomes in Southeastern Nigeria can be achieved.

References

[1] Ezeaku PI, Davidson A. Analytical situations of land degradation and sustainable management strategies in Africa. J. Agri. Soc. Sci. 2008:**4**:42-52

[2] Ezeaku PI, Salau, ES. Indigenous and scientific soil classification systems: A case of differences in criteria in some soils of Northcentral Nigeria. Production Agriculture and Technology Journal:2005:**1**(1):54-66.

[3] Junge B, Deji O, Abaidoo R, Chikoye D and Stahr K. Assessment of past and present soil conservation initiatives in Nigeria, West Africa. Conference on International Agricultural Research for Development; October 9-11, 2007. Tropentag. University of Kassel-Witzenhausen and University of Göttingen; 2007.

[4] Odunze AC. Soil Conservation and Management Practices. Department of Soil Science, IAR/A.B.U Zaria; 2008.

[5] Doe WW, Harmon RS. Introduction to soil erosion and landscape evolution modeling. In: Harmon RS, Doe WW, editors Landscape Erosion and Evolution Modeling. Springer, Boston, MA; 2001. https://doi.org/10.1007/978-1-4615-0575-4_1

[6] Kaffas K, Hrissanthou V. Introductory Chapter: Soil Erosion at a Glance, Soil Erosion- Rainfall Erosivity and Risk Assessment, IntechOpen; 2019. DOI: 10.5772/intechopen.89773. Available from: https://www.intechopen.com/books/soil-erosion-rainfall-erosivity-and-risk-assessment/introductory-chapter-soil-erosion-at-a-glance

[7] Walling DE and Fang D. Recent trends in the suspended sediment loads of the world's rivers. Global Planet Change: 2003:**39**:111-126.

[8] Panagos P *et al.* Global rainfall erosivity assessment based on high-temporal resolution rainfall records. Scientific Reports: 2017:7(4175): 1-11 DOI:10.1038/s41598-017-04282-8

[9] Agwu J. Gender and climate change in Nigeria. A study of four communities in north-central and South-eastern Nigeria. Heinrich Böll Stiftung (HBS); 2009. Available from: www.boellnigeria.org.

[10] Tarolli P, Preti F, Romano N. Terraced landscapes: From an old best practice to a potential hazard for soil degradation due to land abandonment. Anthropocene: 2014. Available from: http://dx.doi.org/10.1016/j.ancene.2014.03.002

[11] Tarolli P, Sofia G. Human topographic signatures and derived geomorphic processes across landscapes, Geomorphology: 2015. Available from: doi:10.1016/j.geomorph.2015.12.007

[12] Nwajide CS. Geology of Nigeria's Sedimentary Basins. CSS Press, Lagos, Nigeria; 2013; 507.

[13] Idike FI. On Appraising Soil Erosion Menace and Control Measures in Southeastern Nigeria Soil Technology: 1992:5:57-65.

[14] Stamp LD. Land utilization and soil erosion in Nigeria. Geographical Review: 1938:**28**:32-45

[15] Skyes RA. History of anti-erosion work in Udi. Farm and Forestry: 1940:**1**:3-6.

[16] Grove A T. Land Use and Soil Conservation in Parts of Onitsha and Owerri Provinces. Geol Surv. Nigeria; 1951; Bull No. 21;74.

[17] Lal R. No-tillage effects in soil properties under different crops in Western Nigeria. Soil Science Society of America Journal:1976:**40**(5):762-768.

[18] Mbagwu JSC, Lal R, Scott TW. Effects of artificial desurfacing on Alfisols and Ultisols in Southern Nigeria: Changes in soil physical properties 1. Soil Science Society of America Journal:1984:**48**(4):1-6.

[19] Ofomata GEK. Soil erosion. Southeastern Nigeria: the view of a geomorphologist, Inaugural lecture series University of Nigeria Nsukka; 1985

[20] Akamigbo FOR. Erosion menace in Anambra state: Government efforts in combating it. A paper presented at the International symposium on erosion in South East Nigeria, Erosion research unit, April 12-14, 1988; Federal university of technology, Owerri.

[21] Nwajide CS. Gullying in the Idemili River catchment, Anambra State, Nigeria: theory and cure. In Freeth SJ, Ofoegbu CO, Onuoha KM, editors. Natural Disasters in West and Central Africa. Vieweg and Sohon Veralagsgesellschaft, Wiesbaden; 1992; 149-162.

[22] Igwe CA. The applicability of SLESMA and USLE erosion models on soils of Southeastern Nigeria [thesis], Department of Soil Science, University of Nigeria, Nsukka; 1994.

[23] Igwe CA. Land use and soil conservation strategies for potentially high erodible soils of central eastern Nigeria. Land Degrad. Develop. 1998:**10**:425-434.

[24] Igwe CA, Akamigbo FOR, Mbagwu JSC. Application of SLESMA and USLE erosion models for potential erosion hazard mapping in South-eastern Nigeria. Int. Agrophysics. 1999:**13**:41-48.

[25] Nwilo P, Olayinka D, Uwadiegwu I,Adzandeh A. An assessment and mapping of gully erosion hazards in Abia State: A GIS approach. Journal of Sustainable Development. 2011:**4**(5):112-122.

[26] Ezemonye MN, Emeribe CN. Rainfall erosivity in Southeastern Nigeria. Ethiopian Journal of Environmental Studies and Management, 2012:**5**(2):112-122

[27] Fagbohun BJ, Anifowose AYB, Odeyemi C, Aladejana OO and Aladeboyeje AI. GIS-based estimation of soil erosion rates and identification of critical areas in Anambra sub-basin, Nigeria. Model Earth Syst. Environ. 2016:**2**(159):1-10

[28] Ibeneme SI, Okoro C, Ezetoha NO, Ahiarakwem CA, Nwankwor GI Integrating Field And Landsat Data In Structural Mapping Of Gully Erosion Sites Within Orlu-Okigwe Axis Southeastern Nigeria. Journal of Earth Sciences and Environmental Studies. 2019:**4**(4):696-707

[29] Ocheli A, Ogbe OB, Aigbadon, OG. Geology and geotechnical investigations of part of the Anambra Basin, Southeastern Nigeria:implication for gully erosion hazards. Environmental Systems Research. 2021:**10**(23)1-16

[30] Odunuga S, Ajijola A, Igwetu N, Adegun O. Land susceptibility to soil erosion in Orashi Catchment, Nnewi South, Anambra State. Proc. IAHS. 2018:**376**:87-95

[31] Chiemelu N, Okeke F, Nwosu K, Ibe C, Ndukwu R, Ugwuoti A. The role of surveying and mapping in erosion management and control: Case study of Omagba erosion site, Onitsha Anambra State, Nigeria. Journal of Environ. and Earth Science. 2013:**3**(11):129-137.

[32] Onweremadu EU. Assessment of mined soils in erosion-degraded farmlands in southeastern Nigeria. Estud. Biol. 2006:**28**(65):59-67.

[33] Lal R. Soil Degradation by Erosion. Land Degradation and Development. 2001:**12**:519-539.

[34] Li Y, Bai X, Tian Y, Lou G. Review and future research directions about major monitoring method of soil erosion. IOP Conf. Ser.: Earth Environ. Sci. 2017:**63**:1-15.

[35] Iwara AI, Ewa EE. Seasonal variability in runoff, sediment and nutrient losses in vegetation fallows in the Rainforest zone of Southern Nigeria. Ghana Journal of Geography 2018:**10**(1):42-57

[36] Lawler DM. Some New Development in Erosion Monitoring: The Potential of Optoelectronic Techniques [R]. School of Geography, University of Birmingham Working Paper 47, 44; 1989.

[37] Prosdocimi M, Jordan A, Tarolli P, Keesstra S, Novara A, Cerda A. The effectiveness of barley straw mulch in reducing soil erodibility and surface runoff generation in Mediterranean vineyards. Science of the Total Environment: 2016:**547**:323-330

[38] Casalí J, Giménez R, De Santisteban L, Alvarez-Mozos J, Mena J, Del Valle de Lersundi J. Determination of long-term erosion rates in vineyards of Navarre (Spain) using botanical benchmarks. Catena. 2009:**78**:12-19. http://dx.doi.org/10.1016/j.catena.2009.02.015

[39] Collins AL and Walling DE. Documenting catchment suspended sediment sources: problems, approaches and prospects. Progress in Physical Geography. 2004:**28**(2):159-196.

[40] Junge B, Mabit L, Dercon G, Walling DE, Abaidoo R, Chikoye D, Stahr K. First use of the ^{137}Cs technique in Nigeria for estimating medium-term soil redistribution rates on cultivated farmland. Soil and Tillage Research. 2010:**110**:211-220

[41] Ganasri BP, Ramesh H. Assessment of soil erosion by RUSLE model using remote sensing and GIS-A case study of Nethravathi Basin. Geoscience Frontiers. 2016:7(6):953-961.

[42] Fashae OA, Ubom O, Adeyemi O. Estimating soil loss from soil erosion-induced land degradation in Uyo metropolis, Nigeria using remote sensing techniques and GIS. Ife Research Publications in Geography. 2013:**12**(1&2):120-137.

[43] Obinna CD, Anejionu PC, Nwilo PC, Ebinne ES. Long term assessment and mapping of erosion hotspots in South East Nigeria. FIG Working Week, Environment for Sustainability, Abuja, Nigeria, 6-10 May, TS03B-Remote Sensing for Landuse Planning; 2013;6448;1-19.

[44] Sorooshian S. Parameter estimation, model identification, and model validation: conceptual-type models In Bowles DS, O'Connell PE, editors. Recent Advances in the Modeling of Hydrologic Systems. 1991: 443-467.

[45] Igbokwe JI, Akinyede JO, Dang B, Alaga T, Ono MN, Nnodu VC, Anike LO. Mapping and monitoring of the impact of gully erosion in southeastern Nigeria with satellite remote sensing and geographic information system; The International Archives of the Photogrammetry, Remote Sensing and Spatial Information Sciences. 2008:XXXVII:865-872.http://www.isprs.org/proceedings/XXXVII/congress/8_pdf/9_WG-VIII-9/03.pdf

[46] Renard KG, Freimund JR. Using monthly precipitation data to estimate the R-factor in the revised USLE. Journal of Hydrology. 1994:**157**:287-306.

[47] Ojo-Atere JO, Ogunwale JA, Oluwatosin GA. Fundamentals of Tropical Soil Science. Evans Brothers (Nigeria Publishers) Limited; 2009.

[48] Roose EJ. Use of the Universal Soil Loss Equation to Predict Erosion in West

Africa. Soil Conservation Society of America, Ankeny, Iowa, 1976; 60-74.

[49] Salako FK. Temporal variation of rainfall erosivity in Southern Nigeria, ASSET Series A, 2007:7(1):190-202.

[50] Salako FK. Rainfall variability and kinetic energy in southern Nigeria. Climate Change. 86:151-164.

[51] Obi ME, Ngwu OE. Characterization of the rainfall regime for the prediction of surface runoffand soil loss in southeastern Nigeria. Beitr. Trop.Landwirtsch. Veterin~irmed. 1988:**26**:39-46.

[52] Arnoldus HMJ. An approximation of the rainfall factor in the Universal Soil Loss Equation. In De Boodt M, Gabriels D, editors. Assessment of Erosion. John Wiley and Sons: Chichester; 1980:127-132.

[53] Fournier F. Carte du danger d'erosion en Afrique au sud du Sahara. 1962; CEC-CCTA, Avril.

[54] Das DK. Introductory Soil Science. Kalyani Publishers: New Dehli, India; 2017:749.

[55] Igwe CA. Gully erosion in Southeastern Nigeria: Role of soil properties and environmental factors. In Godone D, Stanchi S, editors. Research on Soil Erosion. IntechOpen. 2012;8:157-171. DOI: 10.5772/51020. Available from: https://www.intechopen.com/books/soil-processes-and-current-trends-in-quality-assessment/development-of-topographic-factor-modeling-for-application-in-soil-erosion-models

[56] Obi ME, Salako FK and Lal R. Relative susceptibility of some southeastern Nigeria soils to erosion. Catena. 1989:**16**:215-222.

[57] Ofomata GEK. Soil Erosion. Nigeria in maps, Eastern States. Ethiope Publishing House, Benin City, Nigeria. 1975.

[58] Singh G and Panda RK. Grid-cell based assessment of soil erosion potential for identification of critical erosion prone areas using USLE, GIS and remote sensing: A case study in the Kapgari watershed, India. International Soil and Water Conservation Research. 2017:5:202-211.

[59] Brady NC and Weil RR. The Nature and Properties of Soils. 13th ed, New dehli: India Pearson Education Inc.; 2002.

[60] Rellini I, Scopesi C, Olivari S, Firpo M, Maerker M. Assessment of soil erosion risk in a typical Mediterranean environment using a high resolution RUSLE approach (Portofino promontory, NW-Italy), Journal of Maps. 2019:**15**(2) 356-362. DOI:10.1080/17445647.20 19.1599452

[61] Morgan RPC. Soil Erosion and Soil Conservation. Essex: Longman; 1995

[62] FAO. Manual on integrated Soil Management and Conservation Practices. Methods of Soil Conservation held at IITA, Ibadan, Nigeria 21 April- 1 May 1997. FAO, Rome, Italy; 2000

[63] Young A. Agroforestry for soil conservation. Exeter: Wheatons; 1989.

Chapter 3

Therapeutic Landscapes: A Natural Weaving of Culture, Health and Land

Abstract

Current concepts of therapeutic landscape combine landscape with principles of holistic health and the interaction of social, affective and material factors. As social tensions widen the gap between the places of emotional retreat and healing from those of everyday sociability, concepts of therapeutic landscape are evolving to reflect society's current values. This chapter examines how cultural place-based values affect and maintain physical, emotional, mental and spiritual health and well-being in the context of a therapeutic landscape. Five case studies from Australasia, Africa, Middle East and Latin America are analysed to understand better the interrelationships between land, culture and health that make an environment therapeutic. The case studies were selected based on their engagement with the cultural traditions of landscape architecture and how the boundaries of these cultural traditions are negotiated within a modern context. The chapter contributes to the knowledge base of landscape architects and academics interested in the role of culture in producing and maintaining therapeutic landscapes by presenting a cross-cultural analysis to illustrate a range of strategies for incorporating cultural traditions and customs into modern landscape architectural contexts to promote health and well-being.

Keywords: therapeutic landscapes, therapeutic environments, culture, health, well-being, Indigenous knowledge, landscape, landscape architecture

1. Introduction

It has always been known that humans find reassurance in nature. Therapeutic sanctuaries have been sought out through the ages by individuals who have a sense of disorientation or likely future ambiguity. For example, those diagnosed with acute and terminal illness often search for a space of solitude, needing time and space for reflection on significant life and health changes [1]. Explanations for this phenomenon find that the brain is capable of two types of attention: directed attention, belonging to the higher cognitive centres, and soft fascination, linked to the old parts of the brain [2]. In natural environments, the higher cognitive centres of the human brain can rest and reset, lending some of the first scientific evidence to the healing powers of nature.

Most recently, as social tensions associated with the Covid-19 pandemic, climate change and racial inequity deepen the need for places of emotional retreat and healing, therapeutic landscapes have become an increasing topic of interest [3, 4]. However, while more and more research [5–7] finds that contact with nature plays a critical role in psychological well-being, and for those in need, a faster recovery from psychological trauma or stress, Ulrich & Gilpin found when researching hospital gardens, an example that had a measurable negative effect on patient's health [8]. Similarly, following an extensive review of healing gardens, Stigsdotter & Grahn found that not all gardens are healing gardens [9], leading to the question, what is a therapeutic landscape or a healing garden?

The term therapeutic landscape was originally defined in 1992 by the geographer William Gesler as a place "where physical and built environments, social conditions and human perceptions combine to produce an atmosphere which is conducive to healing" [10]. This definition was contested, critiqued and elaborated on over the next two decades, extending from a literal relationship between health and place with the acknowledgment of extraordinary places to a much expanded and refined characterisation. The extended literature reflected the growing interest in the relational and situated approach to well-being, which acknowledged the therapeutic nature of places in the context of those social, cultural, material, affectional and sensual aspects of human and non-human factors [1]. Such therapeutic encounters are defined differently by others, such as 'networks' [11], 'experiences' and 'environments' [12], 'taskscapes' [13], 'mobilities' [14], 'assemblages' [15] and 'enabling places' [16], with the latter gaining particular purchase amongst health and cultural geographers alike.

This theoretical turn saw a much greater emphasis on qualitative and ethnographic methodologies in the study of health designed to reveal the histories, discourses, and lived experiences of a place. Building on the contention that different people experience therapeutic landscapes differently or at different times, the potential healing outcome of the landscape can be seen as a relational process, and therapeutic landscapes become socially and culturally responsive. People's behaviour is deeply embedded within a place, particularly around health [17]. It is not just the space that is healing but the intention of those using the space [18]. Places should not only be defined by the fact they are conducive to healing but also places that are conducive to the maintenance of health and well-being [19, 20].

Successful therapeutic spaces generally reflect a society's current values and aspirations, and as such, become part of its identity. Situated in the current context, 'place' becomes even more challenged and implies that such spaces may need to be flexible to changing interpretation to remain therapeutic over time. Foley & Kistemann [21] considered therapeutic spaces as emergent through a set of embodied experiential practices linking affects, emotions, and bodily sense that arise from being immersed in such therapeutic environments. This form of assemblage can therefore be better understood through its material, metaphorical, and inhabited dimensions [15, 22]. The material component would contain the tangible aspects of landscape that people experience due to their therapeutic qualities. The metaphorical component comprises the ethnographic and cultural values expressed through narratives, myths, and stories that are crucial in defining site-specific rituals and cures. Inhabitation brings together mind, body, and spirit because it draws from lived, experiential, and performative health dimensions.

To this concept, Andrews [17] proposed two streams of application. One relates to the impact of landscape on human experience. The other pertains to how therapeutic landscapes are shaped by the influence of different belief systems, leading to the cultural specificity of the therapeutic landscape concept. In this way, culture and requirements for health create an assemblage of layers that combine in bespoke combinations depending on the individual and the time. The implication is that a healing environment cannot be achieved by a set of design requirements or details you need to meet, it must facilitate a healthcare service that is patient-focused and centres on the diverse needs of all [23]. This extends the therapeutic landscape concept to encompass both tangible and intangible values, supporting the proposition that health, well-being, and place are intricately intertwined and emergent through the layering of architecture and material; with practices and responses in a narrative of individual and communal history where people are simply seeking well-being.

Indigenous methods of restoring health and well-being offer new opportunities for understanding the complexity of health and well-being. Most of the Western understanding around therapeutic landscapes has come from the healing properties of the physical space and adopting a bio-medical approach, translated into spas, mineral springs and mountain retreats [24]. For example, in areas of natural springs, much was made of the unique chemical make-up of the water until scientists uncovered the lack of chemical difference [11]. With this narrow view, what was overlooked was the non-physical dimension to these spaces [24], as scientists failed to engage with and quantify the practices and traditions associated with these landscapes that enabled the healing process [25]. Similarly, what has not been translated into Western culture is the deep-rooted connection with the land and its relation to self-identity on an everyday basis. The associated values and practices embedded in culture and practice are often the foundation of the healing nature of the therapeutic landscapes. Practices in Indigenous cultures are able to be embodied and translated into everyday places and landscapes.

In fact, only a small portion of the literature on therapeutic landscapes discusses the inclusion of Indigenous methods and the relationship between the environment and health and well-being that most of their spiritual connections are grounded by. However, 80% of people in developing countries still rely on traditional medicines or methods. They are inexpensive and easily accessible, they are also believed to be stronger and more effective than treatments offered at a healthcare facility [26]. Indigenous methods and practices often combined with everyday health and well-being, mainly when connected to a sense of place and the sense of identity and symbolic healing the environment can enable [24].

This chapter explores therapeutic landscapes and landscape architectural approaches that can be applied to creating or understanding therapeutic landscapes that are culturally and socially responsive. It addresses the gap in the literature regarding therapeutic landscapes to understand how different social and cultural contexts can affect physical, mental, emotional, and spiritual health. Our intent is to illustrate landscapes that include cultural practice and Indigenous healing methods to promote well-being at both individual and community levels. Examining a series of case studies from around the world, it explores landscape architectural responses to healing. The case studies have been selected based on their success in catering to the social, mental, physical and emotional needs of the user. The chapter unpacks the social and cultural contexts that have shaped these environments and discusses how this can be applied to the universal design of everyday therapeutic landscapes.

2. Method

To investigate the ideas underlying therapeutic environments, a literature review was undertaken using electronic databases such as Scopus, Google Scholar and ProQuest, as well as landmark book publications. Search terms included therapeutic landscapes, therapeutic environments, healthcare landscapes, and restorative outdoor spaces to identify design criteria. Following the first selection and exclusion of non-relevant materials, twenty-two peer-reviewed publications were evaluated.

The articles were critically assessed by conducting a strengths and weaknesses analysis of each study and considering their relevance in promoting health and well-being. As causal relationships between therapeutic environments and human health are difficult to establish, this critical literature review focussed on studies that dealt with association rather than causation. To better understand the theoretical constructs and obtain an international perspective of approaches to health and well-being, 32 case studies were initially selected during the scoping of the literature to demonstrate a breadth of scales. To get an understanding of how different places with differing social and cultural contexts responded through their therapeutic environments, samples were selected from five different countries, namely Kopupaka Reserve in Aotearoa-New Zealand, Lions Park in Queensland, Australia, Freedom Park in Pretoria, South Africa, Jiyan Healing Garden, in Chamchamal, Iraq, and Jardín Etnobotánico de Oaxaca in Mexico, for closer investigation.

3. Kopupaka reserve

Kopupaka Reserve is a large urban wetland in Auckland, New Zealand. It is unique as it combines an expansive natural environment, which has been identified as having unique healing qualities, with Indigenous Māori cultural values. Māori call themselves the *tāngata whenua* (people of the land), which places importance on the connection between land and sea and respect and preservation of the natural environment and its people [27]. Māori collective cultural orientations brings with them a much more holistic approach to health and well-being, combining sustainable and restorative methods with genealogy (*whakapapa*).

In an effort to enhance the presence, visibility, and participation with design, a set of design principles for the creation of designed environments were developed based on Māori cultural values [28]. The values that underpin these design framework are *rangatiratanga* (chieftainship), *kotahitanga* (unity), *kaitiakitanga* (guardianship), *wairuatanga* (spirituality), *manaakitanga* (hospitality), *whanaungatanga* (relationship, sense of family connection), and *mātauranga* (knowledge and wisdom) [29]. Seven principles can be applied to all aspects of design thinking and require an ongoing engagement with Māori tribes. These principles are: *mana* (status of *iwi*, tribe, and *hapū*, sub-tribe, is recognised and respected), *whakapapa* and *mahi toi* (Māori names are celebrated by the recognition of traditional place names on signage and wayfinding and cultural landmarks are acknowledged), *taiao* and *mauri-tu* (the natural environment is protected, restored and/or enhanced), and *mahi toi* (Māori narratives are captured and expressed creatively and appropriately) [28].

The Auckland City Council commissioned the Kopupaka Reserve as a new public space for the emerging Westgate shopping centre. The Kopupaka Reseve is an ancestral name for the area and relates to the meeting of the Tōtara Creek and Waiteputa Stream. The reserve is of a linear shape that follows the riparian corridor of Tōtara Creek and Waiteputa Stream, which had become extremely degraded from the area's long history of growing market produce and most recently a large-scale strawberry farm. Once rich with *māhinga kai* (traditional Māori foods), the waterways had become overloaded with nutrients and infested with weeds. The design restored the native species that once would have grown in the area, including harakeke flax, traditionally used for weaving [28].

Wetland waterways provide an abundance of life with food and purification, contributing to the well-being of the people as Māori maintain that a healthy landscape makes for a healthy individual. Views over the wetlands surrounding the confluence of the two freshwater streams, display the varied birdlife and biologically diverse horticulture. Hard landscape forms are inspired by the abundance of *kai* (food) found in wetlands and guide the boardwalk design (**Figure 1**). Open spaces for communal activities are contrasted with private places for emotional retreat (**Figure 2**). The body is engaged through sitting or walking, the senses are engaged through sights, sound and smells. The park is a celebration of the inseparable bond of water and earth. This is deemed vital to sustaining and balancing the natural environment and which is reflected through wairua and which refers to the spiritual plane. Healthy water and rich resources are deeply embedded in Māori cultural and spiritual landscape values as water is sacred.

A distinct and defining feature of the park is a series of interlocked curving timber structures that frame the wetland ponds and a woven together with vegetated paths. The timber frames are a symbolic representation of the *hīnaki waharua* (eel baskets) that were used to collect the *tuna* (eel) that gathered at one of the historical wetlands in the area [28].

What makes Kopupaka Reserve a successful therapeutic landscape its ability to encourage community engagement and establish a strong sense of place and place identity. These themes are threaded throughout most of the literature that has been written on therapeutic landscapes. As research shifted from its focus on the big event or extraordinary healing spaces to the everyday, the method of healing also shifted from an internal singular process to a communal and inclusive process. Urban park and community gardens emerged as having healing qualities and were celebrated for their unique ability to bring communities together and improve overall well-being. The 22-hectare reserve is located within the centre of Westgate, a new town centre that has emerged as a result of urban sprawl. The park connects together the surrounding suburbs and is a central point for many people. While people are not directly forced to engage with each other or participate in rehabilitative activities, the opportunity is given. Signs and cultural motifs placed around the park tell a narrative of the history of the reserve, people feel more attached to the park as they can understand its significance. The diversity of spaces allows locals to find what they need from the park and offers opportunities to improve their personal health and well-being. The park aimed to restore the mauri (lifeforce) of the area, which was mainly achieved through improved storm water management and the revegetation of key areas improving the water quality of the area and

Figure 1.
Kopupaka reserve and the wetlands, where timber frames are deeply inspired in traditional Māori eel baskets.

Figure 2.
Recreational activities at Kopupapa reserve.

enhancing habitats and ecological corridors [28]. Kopupaka park demonstrates how urban growth and community engagement can be successfully paired with ecology and engineering [30].

4. Lions Park

Lions Park is a small urban park located in Gladstone, Queensland, but in spite of its size, it is considered a revolutionary park for Australia, being one of the first intergenerational and therapeutic environments located within the public realm. It is an inclusive space that has been established for intergenerational socialisation and as a safe space for those with disabilities. Building from the evidence that a strong link exists between the natural environment and improved health and well-being, these spaces provide moments of relaxation, social interaction and physical activity [31]. The park is a destination for locals and it also acts as a thoroughfare between the commercial zone, suburbs and natural environment. The Leonardo Da Vinci inspired playspace incorporates sensory features that can accommodate children and adults with disabilities. The designer, Playscape Creations, worked closely with local communities and those with impairments to create a space that accommodates a diverse range of needs [32]. Due to the urban location, Crime Prevention Through Environmental Design (CPTED) design principles were used throughout the entire design to enhance active and passive surveillance as well as making sure the park was a multi-use space. The increased surveillance also minimises the complexity of maintenance and ensures the park features last longer [32].

The form of the park is inspired by Da Vinci's cogs of interaction. The cogs create islands of play and recreation linked together with wide wheelchair-friendly paths and tactile water features. Each cog is unique with special elements that engage, inspire, and incorporate user preferences for each impairment. For example, those with sensory processing disorders like to spin and swing, but also enjoy shared play and like spaces to hide away in and just observe from. Those with auditory processing disorders engage well with tactile and visual elements but prefer less busy spaces and bold signage. Those with visual impairments are especially sensitive to light so shade is needed along with tactile environments that offer a contrast for navigation. Those with Down syndrome require muscle stimulating equipment, individual and group play opportunities and vestibular play elements. Those with autism can be sensitive to change, so enjoy tactile mapping and colour coding, as well as repetitive play equipment and spaces they can hide away in. For muscular dystrophy, spaces are required to provide shelter from light and heat and have a handhold that can be used with a closed fist. Those with cerebral palsy need rest spaces, handholds that can be used with a closed fist and tactile interactive panels. To accommodate spina bifida, wide wheelchair friendly pathways and transfer points to get on equipment are necessary but also are places for social rest.

Cog one is the entry node at the beginning of the park, which also contains the car park. The purpose of this area is predominantly a gathering space with arranged seating and shelter. From there, the user can move through into the body of the park. Cog two is the linking cog and social hub, connecting either to the nature cog and action cog or the commercial zone cog. In cog three, there is play equipment, informal water play, soft fall mats, turf, and walls that are intended to double as informal seating and shelters. Cog four is the swing, spin and discovery zone and a five-way swing stands in the centre. The play zone is enclosed by a dense vegetation buffer due to its proximity to the road. Cog five is an informal cog dedicated to nature play and exploration play (**Figure 3**). Activities are less structured in this area and there is more opportunity for imagination to take control [33].

In Lions park, play elements and natural features are thereby designed for all of the seven senses; hearing, sight, smell, touch, vestibular and proprioception. Specific play equipment has been included in the design of the playgrounds to ensure that users of all abilities have been included. These play elements include tactile water features for formal and informal water play, small and large shades for those with heat and light sensitivity, safe havens and small hideaways for those who want to escape, hand hold-on equipment that can be used with a closed fist for those who have limited grip, and spinning and full-body vestibular play features. In addition, consideration has been given to stimulating cognitive, social, psychological and physical senses while in the act of playing on a playground [34]. This playground accommodates a variety of bodily movements that stimulate the mind and body in much the same way as the activity of gardening.

Figure 3.
Lions park and the different cog areas.

5. Freedom Park

Freedom Park is located in one of South Africa's capital cities Pretoria. Situated on the top of a hill overlooking the city of Salvokop, the park was mandated by President Nelson Mandela to communicate a narrative of South Africa's journey to freedom following the end of the apartheid. It is a part of the rebuild seeking to address social fragmentation between races and significant disparities in access to education, health care and employment [35]. Freedom Park is a therapeutic landscape in the form of an urban park that tells a narrative of South Africa's journey to freedom and celebrates the 3.6 billion years of physical change in environment as well as those who fought for its freedom [36].

Freedom Park addresses the physical, mental, spiritual, societal and environmental elements of the landscapes of South Africa and embodies them into one space for all. The park supports South Africa's newfound identity, attempts to create new social networks, and supports the once fractured society. It acknowledges the holistic approach of traditional African culture focusing on the collective and bringing community interests to the forefront of any decisions [37]. One African principle that was not lost with colonialism was 'ubuntu', meaning that 'I am what I am because of who we all are', which follows on from the Zulu saying 'umuntu ngumuntu ngabantu', which means 'a person is a person through other persons' [37]. This principle values people first and foremost with a view of the collective whole [38].

One of the most vital features of the park is its engagement to a sense of place. The development of a sense of place is a social and cultural process that depends upon the user's response and cannot be derived from location alone [39]. Creating a sense of place is affirmed in Freedom Park by the inclusion of the historical narrative and the use of traditional practices. When entering through the //Hapo external space the users make their way through traditional and native medicinal plants. The journey educates the user about traditional practices and sparks conversation (**Figure 4**). Sense of place is not only an embedded relationship with the natural

Figure 4.
Use of traditional materials.

environment but directly links to the people within the place [39]. The inclusion of the eleven languages around the park aids in establishing place identity. For example, the //Hapo is from the Khoisan language, S'Khumbuto (the memorial) is Swati, Uitspanplek (A picnic area on the northern side of the hill) is an Afrikaans word, and Isivivane is a Zulu word meaning 'to throw your stone upon the pile' [36].

//Hapo is the first point of arrival to the park. This space is an interactive exhibition telling the story of South Africa over the course of 3.6 billion years. The name //Hapo has been drawn from the phrase '//hapo ge//hapo tama/haohasib dis tamas ka i bo', which means 'a dream is not a dream until it is shared by the entire community'. Within //Hapo there are three external spaces. The first space is the Garden of Indigenous Knowledge/the Healing Garden. In this garden, water channels lead visitors into a quiet and contemplative space where they are immersed in medicinal plants and gain knowledge of Indigenous healing methods. The second space is Sentihaga, the children's area. The landscape has been terraced with a series of cascading walls. Children and others are able to play with water features and plant material. Totems, an amphitheatre and small passageways are features of the garden to spark imagination. The third space is called the Boulders. They are metamorphosed layers of rock composed in a circular formation and surrounded by savannah vegetation to tell the story of South Africa's creation. Indigenous vegetation surrounds //Hapo blending the building into the landscape and preserving the natural terrain of the hillside (**Figure 5**) [36].

Exiting the //Hapo, visitors continue on a contemplative journey up the Salvokop hill along a pathway named the Vhuwaelo. Along the Vhuwaelo there are a series of small gardens that you can weave in and out of, exploring the small private spaces comprised mainly of natural materials and plants. Dramatic views overlook the city of Pretoria and the Union Buildings. At the final moment before you descend down the hill is a space named the Mveledzo. It is made from heavy stones and concrete, providing a final resting spot. Then the visitor descends the hill to arrive at the S'khumbuto, a Swati word for remembering or memorial. The memorial is for those who fought for freedom and showed great leadership. Water has

Figure 5.
Layers of rock surrounded by the savannah vegetation.

been used throughout the memorial because of its importance to African culture and its relation to healing and purification [36].

6. Jiyan healing garden

The Jiyan Healing Garden is a treatment facility and therapeutic garden that was established by the Jiyan Foundation for Human Rights in Chamchamal, Iraq in 2016. The Jiyan Foundation began its work in 2005 by building rehabilitation centres throughout Kurdistan and Iraq for those who were survivors of torture. The healing garden is an animal-assisted trauma therapy centre for those who suffered at the hands of Saddam Hussein's regime and, more recently ISIS. Many children in Iraq suffer from problems associated with this trauma, including behavioural and speech disorders.

A crumbling public infrastructure, including the water and sanitation network, results from years of political unrest. The Jiyan Foundation for Human rights wanted to create a space that was self-sustaining and supported the surrounding environment. The team of architects worked in collaboration with BORDA (Bremen Overseas Research and Development Association) to create a decentralised wastewater system for the garden [40]. This was deemed necessary as the Kurdistan region continually suffers from a rainfall deficit and low groundwater levels. The decentralised wastewater treatment system taps into the sewage pipes that run under the facility and is able to clean 100 cubic meters of dirty water each day, enough water to provide the entire garden with water. The water system also processes animal and human waste, turning it into biogas used for heating the therapy rooms in the winter [41].

ZRS, the architects of the treatment facility and healing garden, wanted to create a space that represented trust, identity and well-being. The facility is an example of traditional and sustainable architecture. The buildings are constructed from local clay material that has been historically used in the area (**Figure 6**). The walls are built from the air-dried brick blended with the thin layer of earth-straw mix. The roof is also constructed from the thin earth straw mix. The earth materials have climate cooling properties that concrete and glass cannot provide. The earth keeps

Figure 6.
Illustration of the building's layout and materials used.

the interior space cool during the hot summers and then keeps the interior dry and warm during the wet and sharp winter months [42].

The treatment facility is situated in a recreated traditional village environment typical to that in Chamchamal. In total, there are 11 individual dwellings that provide therapy rooms and quiet spaces that users can retreat to. A courtyard and shade device connect the dwellings together and allow access to the central healing garden and the animal pens. The garden and buildings create a unique space to heal and the garden is full and flourishing compared to the barren terrain of Chamchamal. Those at the facility participate in group-based garden therapy, which gives them a support network and a new way to overcome their ordeal [42]. The treatment facility also allows them to participate in group activities in the garden and direct therapeutic work in the huts. The technique they tend to use with the children is 'mud therapy'. The kids are encouraged to play outside in the garden and in the mud. They are prompted to get dirty and interact with the many animals before coming back inside for a discussion when they feel safe and comfortable and are often more willing to open up (**Figure** 7) [40].

A distinguishing feature of this therapeutic landscape is the animal-assisted therapy. Those who have been subject to trauma caused by acts of war, torture, violence, or displacement have strong feelings of vulnerability. They are often automatically and subconsciously drawn to the animals. They will try to get close to the animals to feel safe and comfortable. It is also said that the practice of petting, brushing, walking and caring for the animals helps give people something else to focus on, increases self-esteem and reduces feelings of anxiety, grief and isolation. The animals introduced to the Jiyan Healing Garden are a mixture of local animals that many would find in the area as well as exotic animals. Those who use the facility have been subject to generations of pent up trauma and psychological abuse and most are having to adjust to a new life away from their home and loved ones.

The Jiyan Healing Garden adopts place-making strategies and familiarisation through structured individual and communal activities making this space an internationally award winning 'therapeutic landscape'. The gardens provide relief and help to those who are suffering trauma from torture and acts of war and it gives the community access to adequate green space [40]. Green space that is inclusive and contains structured activities is crucial for creating stronger and resilient communities. This is the grounding essence of a therapeutic landscape.

Figure 7.
Outdoor spaces and access to interact with animals.

7. Jardín Etnobotánico de Oaxaca

Jardín Etnobotánico de Oaxaca is deemed one of the world's most original public botanic gardens, it is an ethnobotanical garden on the grounds of the San Domingo monastery in Oaxaca Mexico. Mexico is one of the most biologically rich countries globally, and out of the 31 states in Mexico, Oaxaca is the most biodiverse. Oaxaca has a unique botanical history as it contains evidence of plant domestication of squash and corn in the Americas dating back to the beginning of the new world. The ethnobotanical garden has been designed to reflect and celebrate this unique diversity. Each section of the garden displays a different ecology. The northeast section of the garden is dedicated to the tropical forest. Species that have been included in this section are balsa, ceiba and huaie. The east section of the garden reflects the wet regions of Oaxaca, including species of cacao, vanilla and achiote. The west of the garden is dedicated to the dry lands and includes plant varieties; cacti and agaves [43]. A planter bed planted with squash has been raised above the rest of the garden and faces towards Guilá Naquitz [43].

The garden tells a strong narrative of the natural history of Mexico, as well as exploring the cultural and artistic traditions of Oaxaca. During the site excavation, structures more than 400 years old were discovered and incorporated into the final design. A number of other artists, scientists, anthropologists and horticulturalists collaborated on the design of the planting, hardscape and water systems [43]. The use of materials and hardscaping is unique to the garden and emphasises the narrative of Oaxaca's history (**Figure 8**). A distinct feature of the garden is the green-hued pathway that mimics the formation of a step-fret zigzag. This pattern is common in many pieces of pre-Columbian carving and art. Another distinct feature of the garden is a large water fountain that seeps a distinct red water. The fountain is made from the wood of the Montezuma cypress tree and the water has been dyed red from ground up cochineal coating the wood [43].

The designers and director of the garden choose to include minimal signage around the garden to allow for a more seamless aesthetic. Visitors who want to get a richer experience of the garden and learn about the specific varieties of plant

Figure 8.
Use of softscape and hardscape materials typical to Oaxaca.

Figure 9.
Horticultural practices at the site.

species are encouraged to take a 2-hour tour. During this tour, they are taken through the medicinal and ceremonial plants, including mesquite, copal used in incense, alebriies for carving, and agave plants traditionally used to make mezcal. Visitors are also taken through a section of the garden that is dedicated solely to traditional food crops. The crops include hierba de conejo, corn, beans, chepil (the herb used in tamal de chepil), jicama, amaranth, tomatoes, and chia. Many of these plants were domesticated in Oaxaca. Although, there are around 1000 different varieties of plants in the garden, they are all wild except for the traditional food crops (**Figure 9**) [44].

What makes this space a successful therapeutic environment is the strong sense of place that is created by place attachment and narratives. There are many repetitive and empowering cultural and historical motifs that feature in the garden and speak to Oaxaca's history as one of the more important sites of archaeological history for the Americas, showing the first signs of plant domestication [43]. In the 1970s, seeds of squash, bottle gourd, beans, agave, and chilli peppers were all excavated and included in the ethnobotanical garden design [44]. Visitors to the garden and locals can learn about the ancient history of Oaxaca and the rest of the Americas to understand the role the area played in the development of the new world, creating a strong sense of place identity. But what is also successful about this place is that it began as a passion project from the many locals in the community who wanted to see the preservation of art, culture and the environment. This developed into the current ethnobotanical garden that allows the community to explore the relationship between plants and people and work in harmony with the Cultural Centre. Local designers, artists, scientists, horticulturists, and anthropologists all worked together on the design, construction and infrastructure within the garden. As such, this 'community' garden provides unique opportunities to participate in communal activity and interaction.

8. Discussion

Successful therapeutic landscapes address the social and cultural contexts of their location as well as respond to the environmental, mental and physical needs of each user. In defining therapeutic environments as reliant on a series of encounters, networks, and associations, Duff [16] classified these enabling resources into three categories: social, affective, and material. Social resources are grounded in relationships but are also directly linked to place, since they are a product of it and simultaneously enable everyday experiences. Affective resources are the fusion of individual or collective feelings that define or restrict orientations or actions. Material resources characterise the way in which relationships are established through the materiality of place and how they impact our access to goods, services, and information. Therapeutic landscapes can strengthen and improve the overall well-being of an individual, a community or even a population. In the case studies, a variety of processes were used to foster therapeutic environments, bringing together the relational, affective, emotional and cognitive skills that create and maintain social networks and promote meaningful experiences, closely weaving together land, culture and health.

8.1 The land

Landscapes are experienced in different ways by different people. Different cultural orientations may also lead to different experiences of the same space; however, this does not negate its therapeutic qualities. Similarly, the experience of therapeutic landscapes will vary based on previous experiences or attachments. In the literature, sense of place is discussed as being related to the inclusivity of the design and the intention of the user, but it can also refer to the historical connection with the space, which can enrich the 'place identity'. Placemaking enables the restorative and healing abilities of therapeutic landscapes, aiding those seeking the social inclusion or home comfort and attachment that is often formed through activities and the communal associations made [39]. The spaces that were the most successful in creating a holistic healing landscape for the community, were those environments that were welcoming and established a strong sense of place for all users. Hereby, the sense of place in a therapeutic landscape can result from place identity and place attachment and is often felt by a user when they have an emotional or historical connection to that place.

Two case studies, in particular, illustrate how the land grounds the therapeutic nature of the place. The Kopupaka Reserve in Aoteroa-New Zealand guides visitors along constructed boardwalks. The surrounding waterways provide an abundance of life with views over the wetlands and freshwater streams to the varied birdlife and biologically-diverse horticulture. Open spaces for communal activities are contrasted with private places for emotional retreat, providing a human experience that engages with all the senses through sight, sound and smell. With the attention shift from intimate perspectives to distance views, the approach also changes from a very singular and internal healing process to a communal and inclusive one. The design is permeable and flexible, allowing the materials to flex with the movement of the land and using materiality to will enable the ecology to flow.

Freedom Park in South Africa also uses the land to bring people together. One of the strongest features of this park is its engagement with the sense of place. Bringing together formerly alienated groups in a neutral landscape that offers both distant views over a cityscape and intimate pocket spaces, it is concerned with the physical, mental, spiritual, societal and environmental landscapes of South Africa, embodying them into one space for all. Emotions typically run high when the historical injustices of apartheid are situated in a specific place. The park is designed to seamlessly disappear into its natural context as the materiality of the built environment emulates and draws inspiration from the surrounding nature.

8.2 Culture

Therapeutic landscapes can provide an opportunity to embed cultural and traditional practices that do not often get conveyed in everyday contexts. For example, the Jardín Etnobotánico de Oaxaca (Mexico) began as a passion project for many locals in the community who wanted to see the preservation of art, culture and the environment. The garden tells a story about the cultural and artistic traditions of Oaxaca as well as its place in the natural history of Mexico. There are many repetitive and empowering cultural and historical motifs that feature in the garden and speak to Oaxaca's history. From a social perspective, it brings together artists, horticulturalists, scientists and anthropologists as well as the community at large to reflect the purpose of the garden as an exploration of the relationship between plants and people. The use of materials and hardscaping is also unique to the garden and emphasises the specific narrative of Oaxaca's history. In this project, the health of individuals reflects the strength of the community and collective triumphs are celebrated over the personal gains of the individual.

Another important example of a successful therapeutic environment that celebrates cultural values and includes Indigenous and traditional materials and practices within the design of the landscape is the Kopupaka Reserve (New Zealand). Importance was given to creating a place identity and understanding of how the land was used pre-colonisation. The New Zealand case study adopted a series of principles, Te Aranga framework, to ensure a cultural approach was taken to design that encouraged designers to engage with local Māori tribes in all aspects of the design and construction processes and to ensure that plants and materials had been chosen correctly. Cultural values hereby affect and maintain physical, emotional, mental and spiritual health and well-being. Our findings align with those of Andrews [17], whereby a successful therapeutic landscape has been shaped by the influence of a belief system/culture and emotionally engages with human experience. If these are embedded within the design, the environment will inherently respond to the needs of each individual user in a holistic sense and provide the foundation for a more affluent, healthier lifestyle. For many, connection through place or land includes not just the individual but the entire community of beings, living and non-living, putting the need for connection at the forefront.

8.3 Health

The literature review highlighted some clear themes relating to health that are thread throughout the discussion of therapeutic landscapes and what a successful holistic healing space should provide. Therapeutic environments and healing environments need to be designed to the needs of the user; which includes the needs of the marginalised. Two landscapes that are specifically designed for unique human experience are the Jiyan Healing Garden (Iraq) and Lions Park (Australia).

In the Jiyan Healing Garden, the healing garden is designed for victims of trauma and to address the social tensions that had widened the gap between the places of emotional retreat and healing from those of everyday sociability. Here the healing garden serves as an emotional retreat that facilitates an 'extraordinary' healing experience. For this reason, it takes a unique approach and involves access to animal therapy and pays attention to the specific needs of the user. In addition, it offers the chance for patients to work on the upkeep of a garden actively, activities they may have done at home [23]. Participation at this level creates a sense of ownership and sense of place in the facility, improving the productivity of healing and rehabilitation. Wellness and health of a community do not rely solely on the physical gains from gardening but also the mental gains. Feelings of loneliness and isolation can be mitigated by involvement in communal gardening as well as the practice of plant propagation and cultivation [34]. Concerning the material, a diverse range of spaces and activities are essential to the inclusivity of a space. For this garden, the choice of plants was vital, which targeted the user and their preferred activities [23]. While the garden's main intention is to provide relief and help those suffering trauma from torture and acts of war, it also gives the community access to adequate green space. Green spaces that are inclusive and contain structured activities are crucial for creating more robust and resilient communities. This is the grounding essence of a therapeutic landscape.

Similarly, the Lions Park (Queensland, Australia) has been designed for participation and socialisation on a much smaller scale. With its design focus on intergenerational socialisation and inclusion for those with disabilities, interactive areas provide for various body movements that include vestibular and proprioception and sound, sight, smell, and touch. Social inclusion is achieved through the creation of spaces that meet the needs and desires of all potential users. There are also spaces of refuge that can be used to hide away, for relaxation or contemplation. Playgrounds provide various bodily movements that stimulate the mind and body, similar to the activity of gardening. In particular, play equipment was included within the design of the playground to ensure that users of all abilities were included. Wheelchair-friendly paths, shelters for those who are light-sensitive, tactile panels, visually stimulating signs and planting combinations all combine to create an inclusive and restoratively productive environment.

9. Conclusion

Our understanding of what makes up a therapeutic landscape is evolving. Social issues including; quality of life, access to health care, social inclusion and supportive environments were all addressed in most literature, but overall, we found a lack of engagement with the important cultural context of the therapeutic landscapes and in particular, the inclusion of Indigenous practices. Our research finds that Indigenous methods can provide a holistic approach to health and well-being in our communities and a finer grained understanding of a therapeutic environment.

Our case studies varied in scale from large and expansive natural landscapes to small manmade interventions, but all shared social inclusion as a recurrent theme and all acknowledged aspects of land, culture and health. The successful therapeutic environments were tied to the inclusivity of the design and the intention of the users, but also the historical connection to enrich the place identity. Most of these spaces have come about as projects as a response to passion and request from the local community or as a response to social or cultural demands. This has resulted in the inclusion of the community's needs in the designed spaces and their physical input in the landscape.

The concept of therapeutic landscapes has evolved with research and theory to reflect society's current values. Most recently, this has shifted with the desire to have healthy and strong communities. A stronger sense of place is considered to create stronger communities and improved community-wide health and well-being. The success of a community cannot be separated from the success of its place, including the natural settings, the local culture and wider surroundings. In this way weaving land, culture and health is essential to the concept of the therapeutic environment.

References

[1] Bell SL, Foley R, Houghton F, Maddrell A, Williams AM. From therapeutic landscapes to healthy spaces, places and practices: A scoping review. Social science & medicine. 2018;196:123-130.

[2] Kaplan R, Kaplan S. The experience of nature: A psychological perspective. Cambridge University Press; 1989. 152 p.

[3] Ugolini F, Massetti L, Calaza-Martínez P, Cariñanos P, Dobbs C, Ostoić SK, Marin AM, Pearlmutter D, Saaroni H, Šaulienė I, Simoneti M. Effects of the COVID-19 pandemic on the use and perceptions of urban green space: An international exploratory study. Urban forestry & urban greening. 2020;56:126888.

[4] Honey-Rosés J, Anguelovski I, Chireh VK, Daher C, Konijnendijk van den Bosch C, Litt JS, Mawani V, McCall MK, Orellana A, Oscilowicz E, Sánchez U. The impact of COVID-19 on public space: an early review of the emerging questions–design, perceptions and inequities. Cities & Health. 2020:1-7.

[5] Thompson CW. Linking landscape and health: The recurring theme. Landscape and urban planning. 2011;99(3-4):187-195.

[6] Marques B, McIntosh J, Hatton W. Haumanu ipukarea, ki uta ki tai:(re) connecting to landscape and reviving the sense of belonging for health and wellbeing. Cities & health. 2018;2(1):82-90.

[7] Marques B, McIntosh J, Kershaw C. Therapeutic environments as a catalyst for health, well-being and social equity. Landscape Research. 2021;19:1-6.

[8] Ulrich R, Gilpin L. Healing arts. Nursing. 1999;4(3):128-133.

[9] Stigsdotter U, Grahn P. What makes a garden a healing garden. Journal of therapeutic Horticulture. 2002;13(2): 60-69.

[10] Gesler WM. Therapeutic landscapes: medical issues in light of the new cultural geography. Social science & medicine. 1992;34(7):735-746.

[11] Smyth F. Medical geography: therapeutic places, spaces and networks. Progress in Human Geography. 2005;29(4):488-495.

[12] Conradson D. Landscape, care and the relational self: therapeutic encounters in rural England. Health & place. 2005;11(4):337-348.

[13] Dunkley CM. A therapeutic taskscape: Theorizing place-making, discipline and care at a camp for troubled youth. Health & place. 2009;15(1):88-96.

[14] Gatrell AC. Therapeutic mobilities: walking and 'steps' to wellbeing and health. Health & place. 2013;22:98-106.

[15] Foley R. Performing health in place: The holy well as a therapeutic assemblage. Health & place. 2011;17(2):470-479.

[16] Duff C. Networks, resources and agencies: On the character and production of enabling places. Health & place. 2011;17(1):149-156.

[17] Andrews, G. Landscapes of Wellbeing. In: Brown T, Andrews G J, Cummins S, Greenhough B, Lewis D, Power A, editors. Health geographies: A critical introduction. London: John Wiley & Sons; 2017. p. 59-74.

[18] Bingley, A. Health, People and Forests. In: Convery I, Corsane G, Davis P, editors. Making Sense of Place:

Multidisciplinary Perspectives. Boydell and Brewer; 2012. p. 107-116.

[19] Khachatourians AK. Therapeutic landscapes: A critical analysis [Doctoral dissertation]. Department of Geography: Simon Fraser University; 2006.

[20] Williams A. Therapeutic landscapes as health promoting places. A companion to health and medical geography. 2010:207-223.

[21] Foley R, Kistemann T. Blue space geographies: Enabling health in place. Health & place. 2015;35:157-165.

[22] Foley R, Wheeler A, Kearns R. Selling the colonial spa town: The contested therapeutic landscapes of Lisdoonvarna and Te Aroha. Irish geography. 2011;44(2-3):151-172.

[23] Mei C. Planting design and its impact on efficacy in therapeutic garden design for dementia patients in long-term care facilities in North Texas [Doctoral dissertation]. The University of Texas at Arlington; 2012.

[24] Wilson K. Therapeutic landscapes and First Nations peoples: an exploration of culture, health and place. Health & place. 2003;9(2):83-93.

[25] Bell SL, Phoenix C, Lovell R, Wheeler BW. Seeking everyday wellbeing: The coast as a therapeutic landscape. Social Science & Medicine. 2015;142:56-67.

[26] Gold CL, Clapp RA. Negotiating health and identity: Lay healing, medicinal plants, and indigenous healthscapes in highland Peru. Latin American Research Review. 2011; 93-111.

[27] Marques B, McIntosh J, Hatton W, Shanahan D. Bicultural landscapes and ecological restoration in the compact city: The case of Zealandia as a sustainable ecosanctuary. Journal of Landscape Architecture. 2019;14(1): 44-53.

[28] Auckland Design Manual. Kopupaka Reserve Wetland Park [Internet]. 2017. Available from: http://content.aucklanddesignmanual.co.nz/resources/case-studies/kopupakareserve/Documents/Kopupaka%20Case%20Study.pdf [Accessed: 2020-11-17]

[29] Jacqueline Paul. Exploring Te Aranga Design Principles in Tāmaki. National Science Challenges Building Better Homes, Towns and Cities Ko ngā wā kāinga hei whakamāhorahora [Internet]. 2017. Available from: https://www.buildingbetter.nz/publications/urban_wellbeing/Paul_2017_exploring_te_aranga_design_principles.pdf [Accessed: 2021-04-12]

[30] Isthmus. Kopupaka Park, Landezine [Internet]. 2017. Available from: http://www.landezine.com/index.php/2017/04/kopupaka-park-by-isthmus-group/ [Accessed: 2021-03-21].

[31] Carter M. Wetlands and health: how do urban wetlands contribute to community wellbeing?. In: Finlayson C, Horwitz P, Weinstein P, editors. Wetlands and human health. Dordrecht: Springer; 2015. p. 149-167.

[32] Playscape Creations. Lions Park, Gladstone [Internet]. 2017. Available from: https://www.playscapecreations.com.au/lions-park-gladstone/ [Accessed: 2021-04-04]

[33] Gladstone Council. Lions Park, Gladstone [Internet]. 2017. Available from: https://www.gladstone.qld.gov.au/parks [Accessed: 2021-04-04].

[34] Ferrini F. Horticultural therapy and its effect on people's health. Advances in Horticultural Science. 2003;77-87.

[35] Curtis N. South Africa: The Politics of Fragmentation. Foreign Affairs. 1972;50(2):283-296.

[36] GREENinc, The Freedom Park by GREENinc, Landezine [Internet]. 2012. Available from: http://landezine.com/index.php/2012/09/the-freedom-park-by-greeninc/ [Accessed: 2021-04-11].

[37] Maluleke MJ. Culture, tradition, custom, law and gender equality. Potchefstroom Electronic Law Journal/ Potchefstroomse Elektroniese Regsblad. 2012;15(1).

[38] Gade CB. The historical development of the written discourses on ubuntu. South African Journal of Philosophy. 2011;30(3):303-329.

[39] Lynch KA. The land tells our story: urban native place-making and implications for wellness [Doctoral dissertation]. Boston University; 2016.

[40] Alex MacDonald. Kurdistan 'healing garden' aims to help Iraqis deal with trauma, Middle East Eye [Internet]. 2017. Available from: https://www.middleeasteye.net/news/kurdistan-healing-garden-aims-help-iraqis-deal-trauma [Accessed: 2021-05-10].

[41] BORDA. Nothern Iraq: building a Healing Garden with a sustainable water supply, BORDA WesCA [Internet]. 2018. Available from: https://borda-wesca.org/healing-garden-providing-a-sustainable-water-supply/ [Accessed: 2021-04-28].

[42] ZRS Architekten Ingenieure. Jiyan Healing Garden / ZRS Architekten Ingenieure, Archdaily [Internet]. 2017. Available from: https://www.archdaily.com/883358/jiyan-healing-garden-zrs-architekten-ingenieure [Accessed: 2021-03-22].

[43] Jeff Spurier. Oaxaca's Ethnobotancial Garden, Garden Design [Internet]. 2011. Available from: https://www.gardendesign.com/mexico/oaxaca-ethnobotanical.html [Accessed: 2021-03-25].

[44] Rebecca Bailey. Jardín Etnobotánico de Oaxaca, Que Pasa Oaxaca [Internet]. 2018. Available from: http://www.quepasaoaxaca.com/jardin-etnobotanico-de-oaxaca/ [Accessed: 2021-03-26].

Chapter 4

Community Attachment and Environmental Stewardship: A Peri-Urban Perspective

Abstract

This Chapter questions the negligence of attachment scholarship in the context of environmental stewardship with a specific focus in peri-urban areas. This Chapter has illuminated the imperatives of considering place attachement as an important factor in realizing environment stewardship in peri-urban areas. Three selected hamlets (*Nzasa, Kisarawe and Pugu-Kibaoni*) constitute the study area. A standard closed-ended questionnaire for assessing the extent of attachment of the community was deployed. Literature review on the other hand was used to map baseline information of the study area including the historical significance of the Pugu and Kazimzumbwi forest reserves. Three attachment attributes were explored;, community knowlegiability levels of the area; level of thoughts and feelings of the area; and the extent of community connection to natural resources in the area. It was revealed that the extent of community connection to the forest reserves are relatively strong. The study revealed considerable contrast on forest knowledgiability levels among men and women in the Pugu and Kazimzumbwi forest reserves. Males are generally revealed to be more knowledgiable of the Pugu and Kazimzumbwi forest reserves as compared to their female counterparts. The study revealed that there was substantial relationship between residence status and the level of thoughts and feelings on the forest reserves. The study has shown that natives have more thoughts and feelings of the present and the future of the forest reserves as compared to those who migrated from other parts of the country. The findings suggest that community attachment is of considerable importance in influencing environmental behavior either positively or negatively. Whilist the empirical evidence are drawn from the peri-urban areas of Pugu and Kazimzumbwi forest reserves of Dar es Salaam city, the message thereoff is representing a broad reality in the peri-urban areas of the Global South. The inclusion of community attachment perspective in negotiating environmental stewardship is advocated for as it might contribute in addressing the growing degradation of natural resources in peri-urban areas which has been increasingly declining.

Keywords: natural resource degradation, global south cities, extent of community connection, knowledgiability levels, thoughts and feelings

1. Introduction

There has been encreasing encroachment of the Pugu and Kazimzumbwi forest reserves in recent years. The encroachment has been alarming to the point of threatening the depletion of the forest reserves in the near future. At the same time, studies on land cover changes at the said forests have shown substantial decrease of land cover over the years [1]. This has led to the substantial decline of forest ecosystem goods and services including decrease of water volume on rivers such as nyeburu, decrease of water level on dams (Minaki), decrease of forest products such as honey, wild fruits; rainfall and temperature variability and decrease in crop yields (*Ibid*). Notably, increasing number of studies have recommended interventions geared at altering livelihood strategies of the forest adjacent community as part of the solution to the problem of forest enchroachment and its associated ill-effects. While the issue of livelihood is crucial in addressing the problem in question there are other underlying behavior related concerns to be taken on board for realizing a more holistic solution. Place attachment is one of the behavior related concept linked to community attitudes and behavior which affects positively or negatively community actions towards their surrounding environment. Notably, place attachment concept has been increasingly applicable both theoretically and empirically in rural settings as opposed to urban and peri-urban contexts [2]. While the amount of time one has stayed in a given locale partly explains the limited attachment research in peri-urban areas, this is an area worth schorlarly attention. The need for urgent attachment scholarship in peri-urban areas owes to the increasing degradation of natural resources especially in the Global South. Peri-urban being inhabited by a mixture of migrant populations and pressured by urban externalities poses unique attachment characteristic features in the context of environmental stewardship unlike the urban counterparts. This chapter therefore questions the negligence of attachment scholarship in the context of environmental stewardship in peri-urban areas. The chapter illuminates the imperatives of attachment scholarship in peri-urban areas by examining the commmunity attachment to the Pugu and Kazimzumbwi forest reserves and so establish weather the increasing encroachment is partly attributed to the extent of attachment.

2. Theoretical perspectives

2.1 Place, place attachment and environmental stewardship

Place concept is an increasingly contested concept across disciplines including positivistic scholars, social constructivism and sychometrics. In a review of place literature Low and Altman [3], established that as space is an integrating concept, there is no systematic theory of place and scholars have increasingly echoed luck of conceptual coherence in place research. The concept of place broadly connote to the subjective experience of embodied human experience in the material world [4]. Morgan argues further that place concept is paradoxical, with meaning that is not readily understood, but difficult to define. While understanding place is a step towards comprehending place attachment, this chapter is about place attachment and its implicature to environmental stewardship in peri-urban areas. The next paragraphs of this Section therefore narrows down to the conception of place attachment and its interlinkages to environmental stewardship behavior.

Place attachement is defined as an attitude of bonding (trust and belonging), mutual concerns and shared values with other members of one's group or locale' [5]. "…. Place attachment may contribute to the formation, maintanance and preservation of the identity of a person, group, or culture. And it may also be that place attachment plays a role in fostering individual, group and cultural self-esteem, self-worth, and self-pride …." [3]. Place attachment is also assumed to be of importance to the community since it facilitates engagement in local affairs [6]. Studies variously justify that place attachment motivates indiviaduals to contribute to civic activity on behalf of one's residential location, in the form of sustainable behavior [7, 8], ecological behaviors [9] or reactions to encroachment of one's territory [10, 11]. On this basis, place attachment serves both the individual and large community [6]. Place attachment has also been correlated to community resilience building processes [12] and fostering local friendship [13].

2.2 Peri-urban in the context of environmental stewardship

Situating the peri-urban concept is largely a daunting endeavor. This owes to its increasingly contested milieu by both academia and development practitioners [14, 15]. This Section reviews the conceptual contestation of peri-urban not in the sense of engaging in the debate of its contestations but for understanding its complex nature towards realizing environmental stewardship. The remainder of this Section therefore provide the contested environment and deployed discussion of the peri-urban concept, thereby its linkage to environmental stewardship and to the broad landscape planning discourse.

Broadly, the peri-urban concept is variously and increasingly viewed and conceptualized across both the geographical and disciplinary niches. Arguments behind the multiple conceptions of the peri-urban include lack of scientific definition Forsyth [15], diversity of engaged disciplinary orientations Thuo [14], the difficulties linked to the delimiting the spatial extent of this dynamic region Brook [16] and the equivocation of the concept itself. Notably, scholars increasingly argue that rural, peri-urban and urban environments operate as a system rather than independent entities [17, 18]. At the same time, peri-urban area is increasingly claimed to constitute the intersection point between urban and rural areas [19, 20]. In nutshell, there are growing yet converging understandings within academia on the lived reality of the diverse and context-laden definitions of the peri-urban concept [21]. Another converging understanding linked to peri-urban conception is the co-existence of urban and rural features within cities and beyond their limits [17, 21]. Notably, inspired by the aforehinted converging understandings, this chapter adopts the conception of peri-urban as defined by Mngumi [22], i.e. peri-urban is a city transitional zone, amalgamating both urban and rural landscape functions and features. In addition to the conceptual definition, the empirical materials discussed herein are drawn from the Pugu and Kazimzumbwi forest reserves at the peri-urban of Dar es Salaam city. Dar es Salaam is among the most rapidly growing cities in Africa and currently the business capital of Tanzania. Similar to other peri-urban areas in the Global South [23], natural resources and/ecosystem services in these peri-urban forest reserves are increasingly claimed to deteriorate [22].

Literature has increasingly established that the quantity and quality of natural resources are deteriorating in peri-urban areas, and particularly so in Sub-Saharan Africa [23, 24]. One explanation behind the decline of natural resources in peri-urban areas is the ongoing urbanization which has increasingly resulted in the expansion of the built environment on ecologically sensitive land, especially in peri-urban areas [25–27]. Onother explanation of the growing degradation of natural resources in peri-urban areas in Sub-Saharan Africa is the rapid urbanization due to poverty [28, 29]. The poverty aspect which is largely prevelant in the Global South context renders it difficult for the community members to inculcate the environmental stewardship habit. This is due to the fact that their livelihoods are intricately connected to the resources in their vicinity and so despite having strong natural attachment to the environment, this is compromised by livelihood hardships and as a result leading to low stewardship to the environment. Notably, there is a host of other contributing factors to the decline of natural resources at the peri-urban areas of Pugu and Kazimzumbwi forest reserves. These include climate change, the increase of anthropogenic activities and the expansion of the city's built area to the peri-urban areas [22].

3. Materials and methods

3.1 Case description

The Pugu and Kazimzumbwi Forest Reserves (PKFR), adjacent to the city of Dar es Salaam in Tanzania (**Figure 1**) forms the area where this study was carried

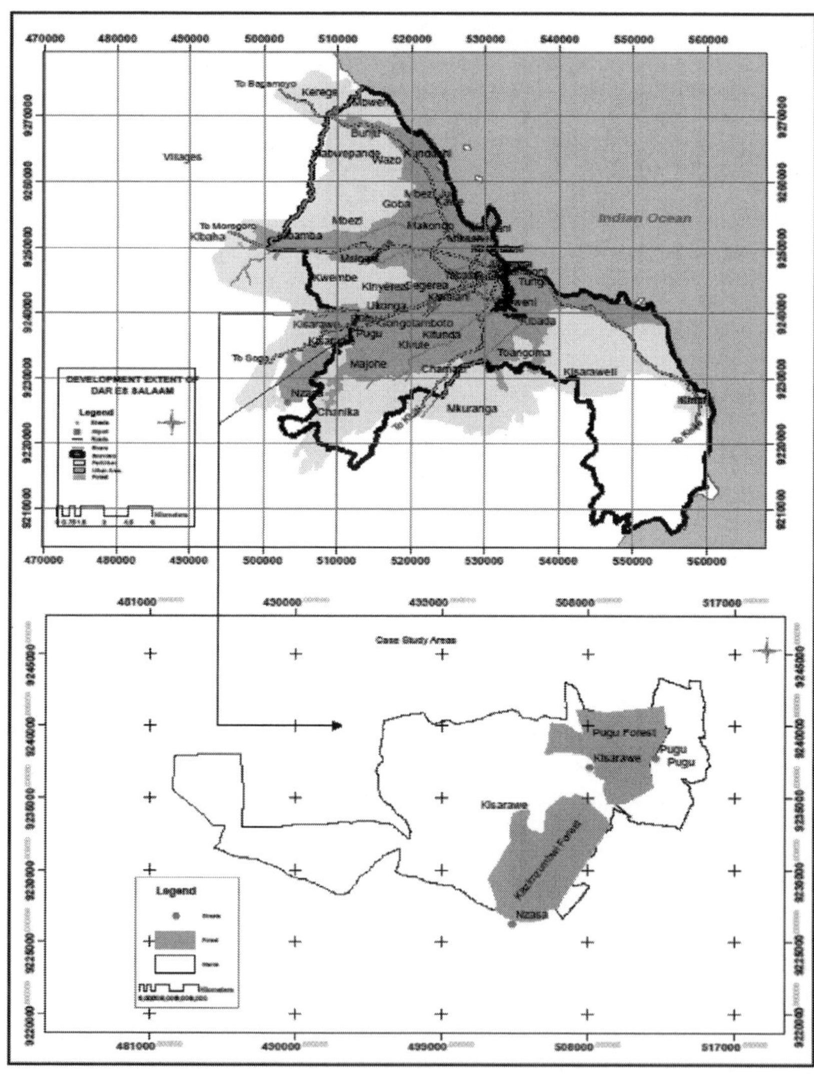

Figure 1.
Pugu and Kazimzumbwi forest reserves, Tanzania. Source: Erimina Massawe, Ardhi university 2019.

out. Administratively, the PKFRs are under the jurisdiction of both districts i.e. Kisarawe district found in Coast region and Ilala district found in Dar es Salaam region; both regions are located along the coastal belt of Tanzania. Historically, the two forest reserves are of global importance as they provide habitats to 37 endemic vertabrate species and about 554 endemic plants [30]. These forests are remnants of few ancient forests in the world, and form the catchment sites for a good number of rivers namely; Msimbazi, Mambizi, Mzumbwi, Vikongoro, Kimani, Nzasa and Nyeburu [1, 31]. Pugu forest gazetted as a forest reserve in the year 1954 lies in the northeastern part of the Pugu Hills, about 25 km southwest of Dar es Salaam and 20 km from the Indian Ocean, and it is positioned adjacent to the Kazimzumbwi Forest Reserve gazetted in the year 1936 [32].

The Indian Ocean tropical monsoon climate is influencing the rainfal and temperature characteristics of both forests. The two forests are characterized by bimodal rainfall pattern i.e. long rains between late March and early June typically known as Masika and short rains between October and December typically known as Vuli. The rainfall in the area i.e. (Pugu and Kazimzumbwi forests) is averaging at approximately 1,100 mm per year whereas Temperature is ranging from 24 to 31° proportional to elevation [32, 33]. In terms of topography, the two forests (Pugu and reserves are positioned between 100 and 305 and 120–280 metres above the sea level respectively [31].

The survey was done in three sub-wards, two of them are located adjacent to the Pugu forest reserve i.e. Pugu-Kibaoni and Kisarawe and the third sub-ward namely; Nzasa is located adjacent to the Kazimzumbwi forest reserve (**Figure 1**). The selection of the three study sub-wards was purposively done so as to capture the possible diversity on attachment discourse in the Pugu and Kazimzumbwi forest reserves.

3.2 Methods

Case study, an imperative approach in characterizing local state of the art on ecosystem services [34] was deployed. Household survey comprise the technique deployed in collecting field data. This was realized through household questionnaire which served as a tool in collecting data on various attributes regarding community attachment to the Pugu and Kazimzumbwi forest reserves and its associated ecosystem services. A standard closed-ended questionnaire for assessing the extent of attachment of the community was deployed. The questionnaire contained 5 score likert scale questions. Literature review on the other hand was used to map baseline information of the study area including the historical significance of the Pugu and Kazimzumbwi forest reserves. A total of fifty (50) questionnaires were randomly administered in each of the study sub-ward amounting to a total of one hundred and fifty (150) in the entire study area.

Interview consent was requested to all participants from the three study sub-wards, this was done so as to be able to proceed with the interview smoothly. This was done so as to comply with ethical requirements. The household interview through structured (close-ended) questionnaire took place after three days training enumerators. The training was purposed at making sure that the main themes of the questionnaire and the specific questions and likely response from the interviewees/respondents are aparently clear among the enumerators. This procedure was crucial as enumerator's work involved asking questions to respondents, providing clarification in case the question (s) becomes not so clear to respondents and thereafter fill the questionnaire after capturing response from interviewees/respondents. This was again important since it ensured maximum response as some respondents could not read and right. Furthermore, at every evening of the field work, a brief assessment with enumerators was held regarding the work of the day including cross checking the way questionnaires have been filled and planning for the day ahead. This was necessary to check the work progress and address the challenges if any for ensuring validity of the data collected. Data analysis was done through SPSS and excel software. The analysis was dominated by descriptive statistics whereby the findings in terms of barcharts and pie charts showing distribution of various attributes on attachment were produced.

4. Results and discussion

This section presents the findings on both respondents characteristics and the findings on community attachment on the Pugu and Kazimzumbwi forest reserves. The section starts by describing the demographic characteristics followed by attachments to the forests reserves of Pugu and Kazimzumbwi.

4.1 Selected participants' characteristics

Participants of this study included 150 randomly selected adults 18 through 82 of age with the mean age of 42.8 and (S.D = 17.9, 58 female); 83 were natives, 51 were migrants from the capital Dar es Salaam and 13 were migrants from upcountry. The participants were included from three sub-wards surrounding the Pugu and Kazimzumbwi forest reserve in order to capture hybrid and diverse attributes on attachment to the reserves. Each sub-ward had 50 participants in the survey. The main characteristics of participants are summarized in **Table 1**.

4.2 Community attachment to forest reserves in Pugu and Kazimzumbwi forest reserves

The aspect of community attachment to the forest reserve were assessed through household survey via a number of questions as detailed in the following section (s).

Respondent's characteristics		Male		Female		Total	
		N = 92	%	N = 58	%	N = 150	%
Education status	Primary or less	67	72.8	53	91.4	120	80
	Secondary	21	22.8	4	6.9	25	16.7
	High school	1	1.08	0	0	1	0.7
	Certificate and above	3	3.3	1	1.72	4	2.7
Residence status	Native	47	51.1	36	62.1	83	53.3
	Migrant from DSM	36	39.1	18	31.0	51	34
	Migrant from upcountry	9	9.7	4	6.8	13	8.6
Geographic location	Nzasa	22	23.9	28	50.9	50	33.3
	Pugu-Kibaoni	36	39	14	25.5	50	33.3
	Kisarawe	34	36.9	16	27.6	50	33.3

Source: Field work, 2018.

Table 1.
Selected characteristics of participants.

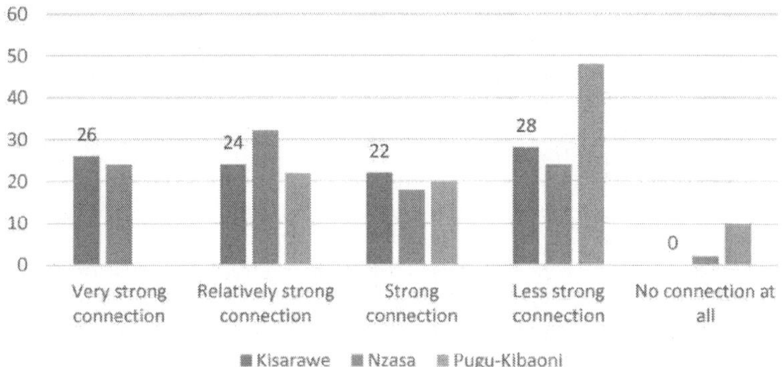

Figure 2.
Extent of connection to the forest reserves (%; N = 150). Source: field work 2018.

4.2.1 Extent of connection to the forest reserves

Figure 2 presents findings on the extent of community connection to the Pugu and Kazimzumbwi forest reserves. Generally; the extent of connection to the forest reserves are more or less similar between Nzasa and Kisarawe sub-wards. About a quarter (1/4) of respondents in these two sub-wards had shown relatively strong connection to the Pugu and Kazimzumbwi forest reserves. At the same time, Pugu-kibaoni sub-ward has outweighed the two sub-wards by having high percentage scores on less strong connection to the reserves. At Pugu-Kibaoni sub-ward, about half (1/2) of the respondents casted their scores on less-strong connection and a tenth (1/10) had their scores on no-connection at all to the forest reserves. The reason for Pugu-Kibaoni having considarable low level of connection to the reserves is due to the fact that the area is located near to the city as compared to the other two sub-wards. This being the case the livelihoods of the majority of its dwellers are not directly linked to the forest reserves as they engage mostly in trade and small business activities. This is backed up by the correlation analysis between the extent of connection to the forest reserves and the extent of one's life attachment to the forest reserves. This had a correlation score of 0.5 testifying that there is a considarable linkage between one's connection to the forest reserves and the connection of an individual's life to the forest reserves.

On the other hand the factors behind community attachment to the forest reserves were gathered during focus groug discussion (FGD) in all the three sub-wards. One of the autspoken reason for having strong connection to the Pugu and Kazimzumbwi forest reserve is the issue of livelihood connection with the forest reserves. During the FGDs it was underscored that a good proportion of the dwellers in areas at the vicinity of the forest reserves owes their living in the same. The activities in the forest reserves that derive a living to forest adjacent communities includes charcoal making and selling, timber business, fire wood fetching/cutting and selling, bee keeping and collecting other forest products such as medicene (herbs) and wild fruits. One participant in a FGD at Kisarawe had these to say ...

> *...I have been depending on the forest for earning a living for almost my entire life. This has made my life to be strongly connected to the forest reserves. This is the reason why I feel like I can not afford to live without these forest reserves. Nowardays there are some restrictions on how to access the reserves but I know how to find my way through the forest without ending up in the hands of those with authorities...*

4.2.2 Knowledge of the Pugu and Kazimzumbwi forest reserves

The question of wheather the community members were knowledgiable of the forest reserves was also asked in the course of establishing the level of community attachment. The findings on knowledgiability levels on Pugu and Kazimzumbwi forest reserves which partly tells on their attachment as as summarized **Figure 3**. The levels of knowledgiability was analyzed against sex of respondents.

The study revealed considerable contrast on forest knowledgiability levels among men and women in the Pugu and Kazimzumbwi forest reserves. Accordingly; males are generally revealed to be more knowledgiable of the Pugu and Kazimzumbwi forest reserves as compared to their female counterparts (**Figure 3**). Whereas a third (1/3) of female respondents were deduced to be less knowledgiable of the Pugu and Kazimzumbwi forest reserves, about a third (1/3) of males claimed to have complete knowledge of the reserves. The sex difference on knowledgiability on forest reserves also featured during the FGDs. It was explained that most men could get along the forest reserves with easy as compared to men. In a FGD at Nzasa, one female participant in a FGD expressed herself this way...

> *...In this forest reserve (Kzimzumbwi forest resrve) although both male and female access the reserves for gendered distinguised interests, males are more familier with the forest reserve. This is partly because men spends more time in the forest extracting resources for the living as compared to women....*

The variable of knowledgiability of the forest reserves was further correlated with the variable ' I know where all the paths leads to'. The correlation score for the two variables was 0.8 signifying that there were strong relationship between general knowledgiability of the forest reserves and the knowledge of where paths leads to in the forest reserve.

4.2.3 Thoughts about the present and the future of the forest reserves

Thoughts and feelings about the present and future of something tells the extent at which one is linked and/attached to the same. In this regard, the feelings and/thoughts about the present and the future of the Pugu and Kazimzumbwi forest reserves by the adjacent communities were examined. The findings on

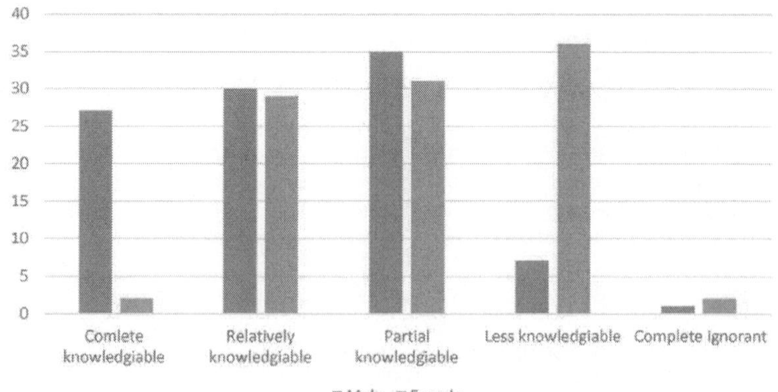

Figure 3.
Knowledge of the forest reserves by the community (%: N = 150). Source: Field work 2018.

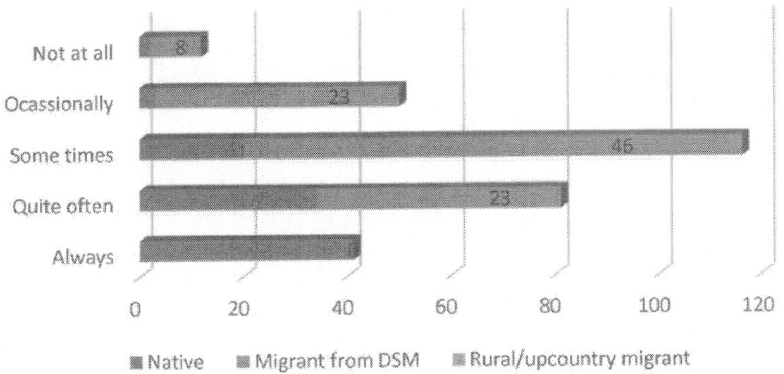

Figure 4.
Present and future thoughts about the forest reserves (%: N = 150). Source: field work 2018.

Figure 4 summarizes the stutus-quo on thoughts and feelings about the present and the future of the forest reserves.

The analysis on community thoughts and feelings about the present and the future of the forest reserves were done against the residence status.The study revealed that there was substantial relationship between residence status and the level of thoughts and feelings on the forest reserves. In nutshell, the study has shown that natives have more thoughts and feelings of the present and the future of the forest reserves as compared to those who migrated from other parts of the country. More than a third (1/3) of the natives claimed to *always* think about the present and the future of the forest reserves and there was non of the migrants who always had a thought about the present or the future of the forest reserves. On the other hand about half of the migrants from Dar es Salaam and those from upcountry had *some times* thought regarding the present and the future of the Pugu and Kazimzumbwi forest reserves. These findings on thoughts of the present and the future of the Pugu and Kazimzumbwi forest reserves suggests that the more one stays with a natural resource the more one becomes with feelings and thoughts about their present and their future prospects. This in turn tells on the extent at which one is attached to the particular resource and in this case the Pugu and Kazimzumbwi forest reserves.

5. Conclusion

This chapter questions the negligence of attachment scholarship in the context of environmental stewardship with a specific focus in peri-urban areas. Peri-urban being inhabited by a mixture of migrant populations and pressured by urban externalities poses unique attachment characteristic features in the context of environmental stewardship unlike the urban counterparts. This chapter has illuminated the imperatives of considering place attachement as an important factor in realizing environment stewardship in peri-urban areas. Whilist the empirical evidence are drawn from the peri-urban areas of Pugu and Kazimzumbwi forest reserves of Dar es Salaam city, the message thereoff is representing a broad reality in the peri-urban areas of the Global South. The inclusion of community attachment perspective in negotiating environmental stewardship is advocated for as it might contribute in addressing the growing degradation of natural resources in peri-urban areas which has been increasingly declining. Researchers and policy makers alike should take this up by including a focus on community attachment in the course of forging environmental stewardship in these transitional spaces in the Global South. This will in turn contribute to increase environmental welfare and reduce the growing degradation of natural resources in peri-urban areas.

References

[1] Lupala, J.M., M.V. Mdemu, and S.P. Butungo, Effects of peri-urban land use changes on forest ecosystem services: the case of settlements surrounding Pugu and Kazimzumbwi Forest Reserves in Tanzania. Journal of Geography and Geology, 2014. 6(4): p. 231.

[2] Raymond, C.M., G. Brown, and D. Weber, The measurement of place attachment: Personal, community, and environmental connections. Journal of environmental psychology, 2010. 30(4): p. 422-434.

[3] Low, S.M. and I. Altman, Place attachment, in Place attachment. 1992, Springer. p. 1-12.

[4] Morgan, P., Towards a developmental theory of place attachment. Journal of environmental psychology, 2010. 30(1): p. 11-22.

[5] Perkins, D.D., J. Hughey, and P.W. Speer, Community psychology perspectives on social capital theory and community development practice. Community Development, 2002. 33(1): p. 33-52.

[6] Lewicka, M., Ways to make people active: The role of place attachment, cultural capital, and neighborhood ties. Journal of environmental psychology, 2005. 25(4): p. 381-395.

[7] Uzzell, D., E. Pol, and D. Badenas, Place identification, social cohesion, and enviornmental sustainability. Environment and behavior, 2002. 34(1): p. 26-53.

[8] Guàrdia, J. and E. Pol, A critical study of theoretical models of sustainability through structural equation systems. Environment and Behavior, 2002. 34(1): p. 137-149.

[9] Vorkinn, M. and H. Riese, Environmental concern in a local context: The significance of place attachment. Environment and behavior, 2001. 33(2): p. 249-263.

[10] Kyle, G., et al., Effects of place attachment on users' perceptions of social and environmental conditions in a natural setting. Journal of environmental psychology, 2004. 24(2): p. 213-225.

[11] Warin, M., et al., The power of place: space and time in women's and community health centres in South Australia. Social Science & Medicine, 2000. 50(12): p. 1863-1875.

[12] García, I., F. Giuliani, and E. Wiesenfeld, Community and sense of community: The case of an urban barrio in Caracas. Journal of Community Psychology, 1999. 27(6): p. 727-740.

[13] Perkins, D.D., et al., Participation and the social and physical environment of residential blocks: Crime and community context. American journal of community psychology, 1990. 18(1): p. 83-115.

[14] Thuo, A.D.M., Unsettled Settled Spaces: Searching for a Theoretical 'Home'for Rural-Urban Fringes. International Journal of Scientific and Research Publications, 2013. . 3(7).

[15] Forsyth, A., Defining Suburbs. Journal of Planning Literature, 2012. 27(3): p. 270-281.

[16] Brook, R.M., The Peri-Urban Interface Hubli-Dharwad, India Changing Frontiers. Books for Change. 2001, Bangalore, India.

[17] Allen, A.D., and Hofmann P, Governance of Water and Sanitation Services for the Peri-urban Poor; A Framework for Understanding and Action in Metropolitan Regions, U. Development Planning Unit, United Kingdom, 2006

[18] Wandl, A. and M. Magoni, Sustainable Planning of Peri-Urban Areas: Introduction to the Special Issue. Planning Practice & Research, 2016. 32(1): p. 1-3.

[19] Olujimi, J. and K. Gbadamosi, Urbanisation of Peri-Urban Settlements: A Case Study of Aba-Oyo in Akure, Nigeria, The Social Sciences, 2007. 2(1): p. 60-69.

[20] Birkmann, J., et al., Adaptive urban governance: new challenges for the second generation of urban adaptation strategies to climate change. Sustainability Science, 2010. 5(2): p. 185-206.

[21] Salem, M., Peri-urban dynamics and land-use planning for the Greater Cairo Region in Egypt. Sustainable Development, 2015. 168: p. 1109.

[22] Mngumi, L., Climate change resilience: exploring socio-ecological system resilience for livelihood effects of climate change in peri-urban areas. 2021.

[23] Roy, M.K., et al., Climate change and declining levels of green structures: Life in informal settlements of Dar es Salaam, Tanzania. Landscape and Urban Planning, 2017.

[24] MEA, Ecosystem and human well-being: biodiversity synthesis. World Resources Institute, Washington, DC, 2005.

[25] Kombe, W.J., Land use dynamics in peri-urban areas and their implications on the urban growth and form: the case of Dar es Salaam, Tanzania, Habitat International, 2005. 29(1): p. 113-135.

[26] Marshall, F., Waldman, L., MacGregor, H., Mehta, L. and Randhawa, P., On the Edge of Sustainability: Perspectives on Peri-urban Dynamics, STEPS,. 2009.

[27] Radford, K.G. and P. James, Changes in the value of ecosystem services along a rural–urban gradient: A case study of Greater Manchester, UK, Landscape and Urban Planning, 2013. 109(1): p. 117-127.

[28] Kestemont, B., L. Frendo, and E. Zaccai, Indicators of the impacts of development on environment: A comparison of Africa and Europe. Ecological indicators, 2011. 11(3): p. 848-856.

[29] Niemelä, J., et al., Using the ecosystem services approach for better planning and conservation of urban green spaces: a Finland case study. Biodiversity and Conservation, 2010. 19(11): p. 3225-3243.

[30] Burgess, N.D., Global importance and patterns in the distribution of coastal forest species, in Coastal Forests of East Africa. 2000. p. 235-248.

[31] TFCG, Two surveys of the plants, birds and forest condition of Pugu and Kazimzumbwi Forest Reserves in 2011 and 2012, T.T.P. 36, 2013.

[32] Clarke, G. and A. Dickinson, Status Reports for 11 Coastal Forests in Coast Region, Tanzania Frontier-Tanzania Technical Report No. 17. The Society for, 1995.

[33] Lupala, J. and C. Maglan, Climate change and its effects on livelihood strategies of peri-urban coastal communities in Tanzania. 2015.

[34] Lamarque, P., et al., Stakeholder perceptions of grassland ecosystem services in relation to knowledge on soil fertility and biodiversity. Regional environmental change, 2011. 11(4): p. 791-804.

Chapter 5

Contributing to Healthy Landscapes by Sustainable Land Use Planning: A Vision for Restoring the Degraded Landscape of the Centre Region of Portugal

Abstract

The ecological-based methodologies are determinant to develop complete strategies in restoring the ecosystems at a landscape scale. Those methodologies start with comprehending ecological processes by mapping fundamental structures of the territory (water, soil, biodiversity), also called green infrastructures. The adequate land use planning and its forthcoming implementation will guarantee a multifunctional landscape, better ecosystem services provision, and a possibility of developing new economies. The intervention of Landscape Architecture at the landscape scale will also provide information about the place and the type of restoration actions to be implemented. The Centre Region was the most affected by rural fires from 2017, representing 15% of the total region area (416 thousand hectares). These events reflect the high importance of rethinking the territory with more suitable land uses, considering the concepts of sustainability, resilience, and ecological integrity. This work proposes a Landscape Transformation Plan for the Centre Region of Portugal, applying the FIRELAN model. The results show that about 35% of the Centre Region should have restoration action towards a more sustainable landscape.

Keywords: fire resilience, green infrastructure, ecosystem restoration, landscape transformation

1. Introduction

The first Portuguese Landscape Architect, Francisco Caldeira Cabral, reflected on how nature conservation should not be seen from a museology perspective, where Man is external to the object of protection. He defended that every person is an integral part of nature conservation by actively participating in protecting natural resources and constructing the landscape.

> *(...) a campaign of general mentalization began for the need as a condition of urban and rural life, to maintain in congruent form the essential elements of the*

> *natural landscape, conserving or even reconstituting its continuity and functionality. Thus became aware of maintaining the " continuum natural " and the " continuum cultural [1].*

Since our existence, humankind has established interactive and empirical relationships with the Landscape [2], searching for defensive systems at higher elevations and safeguarding the fertile valleys as food producers essential for survival. But the interaction between Man and the ecosystems is bidirectional, where an action of Man on a particular ecosystem will imply a reaction and an adaptation of the ecosystem [3].

Landscape and land-use planning are intended to plan human interventions that maintain or promote the landscape's dynamic stability. The stability of the landscape is associated with slow landscape evolutions (pedogenesis), while instability is characterized by rapid changes (morphogenesis) [4]. The balance between morphogenetic and pedogenetic processes is a natural process of the landscape, which can be intensified towards instability by an incorrect action in the territory. The planning of human intervention in the territory in harmony with ecological systems has resulted in preserving natural resources and nature conservation.

The understanding of the functioning of natural systems, or ecosystems, as a support for decision making [5] emerged after the foundation of ecology as a science, in the mid-nineteenth century, by Ernest Haeckel. The evolution of knowledge about the ecological processes in a given territory allowed the development of ecology-based landscape planning methodologies.

Following the Convention on Biological Diversity [6], one of the decisions taken at the fifth Conference of the Parties to the Convention on Biological Diversity [7] was that biodiversity conservation objectives could only be achieved through an ecological-based approach:

> *(...) a strategy for the integrated management of land, water, and living resources that promotes conservation and sustainable use in an equitable way. It is based on the application of appropriate scientific methodologies focused on levels of biological organization that encompasses the processes, functions and interactions among organisms and their environment. It recognizes that humans, with their cultural diversity, are an integral part of ecosystems [7].*

Ecological-based methodologies start from the knowledge and spatialization of natural processes that occur in a given territory [8]. With this approach, the most significant areas for ecosystem functioning are identified in the landscape, with various potentialities, and actions are planned without compromising the stability and balance of the landscape. This approach is related to the concept of ecological suitability to the various human activities that do not compromise the proper functioning of ecological processes [9].

The concept of ecological suitability was used in the United States of America and the United Kingdom, using the manual overlay of transparent supports, whose methodologies were refined throughout the 1960s [10]. Other landscape architects followed, such as the work done by Philip Lewis, in 1964, to classify all the environmental resources of the State of Wisconsin, with the purpose of delimiting the areas where building should not be done [11]. The concept of ecological suitability was also considered in McHarg's [12] and Steiner [13] methodologies with the study of environmental processes and cultural integration in the choice of the best use, according to the intrinsic characteristics of the systems. In Portugal, contemporary with McHarg's methodology, the Algarve Plan was completed by landscape architects A. Barreto, A. Castelo-Branco and A. Dentinho, with methodologies based on ecological suitability of the Landscape [11].

The adequate planning of the landscape, according to the ecological processes that occur in it, impacts not only the ecological balance but also on the economic and social balance. This type of intervention has the ability, in a cost–benefit analysis, to be the best way to prevent future costs arising from natural disasters (flood prevention, fire risk reduction, mass movement), maintain water quality, ensure greater agricultural productivity, and contribute to the enhancement of urban areas [14]. Of this last point, the study done [15] in the case study of Cologne (Germany) concluded that increasing urban parks by 1%, about 500 meters from housing, leads to the growth of housing sales prices by 0.1%. With a priori protection of natural resources, engineering solutions are less, and the cost is lower [14].

The economic valuation of ecosystems was developed with the emergence of the ecosystem services concept [16] and spread at the beginning of the 21st century with the publications of the Millennium Ecosystem Assessment. Different local scientific and political communities started raising awareness of the importance and benefits of ecosystem services. Several initiatives and methodologies have emerged to quantify these services in monetary value. However, there are sometimes limitations in quantifying an ecosystem service [17].

Ecological-based planning has the added advantage of helping to increase the number or quality of services provided by ecosystems. Inherent in the concept of ecological-based planning is the continuity and ecological network [18]. In fact, landscape connectivity is a component of landscape structure that facilitates or impedes the flows of natural cycles [19]. Those authors believe that it is more critical to establish connectivity than the proximity of areas studied in "island biogeography" [20].

Like nature conservation, the ecological continuity of ecosystems is a broad concept that involves the need to promote continuity between plant and animal species and that of all ecological cycles: water, nutrients, carbon, etc. For this, the planning of these continuities (ecological structures or networks) cannot be disconnected from the function and its "congruent form", as defined by the landscape architect Francisco Caldeira Cabral [1]. The land-use planning approach advocated, plans the landscape from a biophysical and not only biological perspective [11, 21], where the continuous structures of the landscape are identified and planned according to their function and coherent form.

The importance of connecting nature protection areas, establishing a network or infrastructure has been reinforced [22]. According to this publication, this type of infrastructure, designated by green infrastructure, can mitigate fragmentation and promote the various benefits of maintaining and restoring ecosystems and their services, not only inside Natura 2000 areas.

However, as mentioned above, nature conservation should be understood as a comprehensive and integrative concept of Man and ecological processes, not limited to classified areas (RAMSAR Areas, Biogenetic Reserves, Natura 2000 Network, etc.). Nature conservation will emerge from the correct occupation and use of the Landscape by Man, both in terms of building, forest, woodland and/or agriculture. Therefore, the incorporation of ecological processes in the nature conservation strategy is necessary [23], not forgetting that: (1) ecosystems are spatially and temporally dynamic; (2) ecosystem components interact with each other contributing to biodiversity; (3) ecological processes act as species-selective forces; (4) in highly modified sites ecosystem restoration is a conservation priority.

2. Integration between green infrastructure, ecosystem services, and ecosystem restoration

The concept of structure in landscape architecture is necessarily related to spatialization. In this structure, relationships between ecosystems or elements are expressed [24]. The structure integrates the landscape system's objective and subjective components, articulating significant features that relate to each other [11]. Infrastructure is a structure that serves as the base for something to be developed. Thus, when we refer to green infrastructure, we are talking about a planned network of structural spaces rather than a network of spaces disconnected from the biophysical structure of the territory. The concept of green infrastructure is broad and varied, but it is considered the one summarized by Naumann [25], whose definition is frequently used:

> *Green infrastructure is the network of natural and semi-natural areas, features and green spaces in rural and urban, terrestrial, freshwater, coastal and marine areas, which together enhance ecosystem health and resilience, contribute to biodiversity conservation and benefit human populations through the maintenance and enhancement of ecosystem services. Green infrastructure can be strengthened through strategic and coordinated initiatives that focus on maintaining, restoring, improving and connecting existing areas and features as well as creating new areas and features [25].*

Green infrastructure planning involves an assessment of the types of natural and cultural resources available and a prioritization of the resources most important to present and future needs [14]. Therefore, a green infrastructure strategy includes the process of identifying, assessing, and prioritizing areas that are critical to preserving a healthy community. In addition to prioritizing areas, there is also a need to implement actions to ensure their conservation [14]. Mapping natural resources are thus the first step in building a green infrastructure map to inform which areas need conservation actions and which need restoration actions.

The European Biodiversity Strategy for 2020 [26] highlighted the importance of using green infrastructures in landscape planning since it can ensure the "best functional connectivity between ecosystems within and between Natura 2000 areas and in the wider countryside" ([27]: 6). The indication in a European document of the importance of ensuring functional connectivity between ecosystems, even outside protected areas, through landscape planning was a crucial step towards elevating nature conservation to a more comprehensive status than protecting particular species. Recently, the European Biodiversity Strategy for 2030 [28], as a core part of the European Green Deal, defines an action plan towards protecting nature and reversing the degradation of ecosystems.

Green infrastructure is also a tool to achieve economic and social benefits through natural solutions. This concept of natural solutions (or nature-based solutions) was further developed by the working groups of the European Commission [27] as a solution inspired, supported or copied from nature. The green infrastructure strategy itself states that "Green infrastructure can make a significant contribution to the effective implementation of all policies where some or all of the desired objectives can be achieved in whole or in part through nature-based solutions." ([29]: 3).

The scientific community widely refers to ecosystem services as benefits that a population acquires, directly or indirectly, from ecosystem functions [16]. Ecosystem services result from flows of materials, energy, and information from natural capital stocks capable of producing human well-being [16]. Their monetary quantification [30], with specific units [31] or measured through indicators [32] as well as a qualitative assessment [33] has been addressed in the last decades.

These services were categorized into several typologies by the Millennium Ecosystem Assessment [34]: supporting (services required to produce all other ecosystem services); provisioning (products obtained from ecosystems); regulating (benefits obtained through the regulation of ecosystem processes); and cultural (non-material benefits obtained from ecosystems). The landscape is intended to be one where ecosystem services are provided in balance with the physical structure that supports them, so the various actors dealing with land-use planning must understand the support structure of the landscape.

To contribute to a consistent definition of ecosystem services, Fisher [35] introduces the importance of ecosystems' structure and their processes and functions. In this approach, ecosystem services are characteristics of ecosystems used directly or indirectly by humans to produce well-being. Accordingly, ecosystem structure is itself a service because it provides a platform for ecosystem processes. Related to this idea, the same authors say that the configuration of ecosystem structure and processes is necessary for the healthy functioning of ecosystems and their services, relating them to the concept of (green) infrastructure. The spatial characteristics of ecosystems are also a way to classify their services, so it will be important in planning to know what services are available and how they flow through the landscape. In this way, it is important to understand the relationships between the production of the service and the place where the benefit occurs, recognizing the dynamic characteristics of ecosystems. In this regard, [35] propose a system for classifying ecosystem services into three categories: (1) in situ (services produced and benefits provided occur at the same location); (2) Omni-directional (where services are provided at a single location, but the benefit occurs in the landscape surrounding the service production with no defined direction); (3) directional - where service provision benefits from a specific location due to the direction of flow.

The Ecosystem services were also an integral part of the European Biodiversity Strategy for 2020 [26]. According to this strategy, ecosystems and their services would be maintained and enhanced by creating green infrastructure and restoring at least 15% of degraded ecosystems. Ecosystem services were then identified through indicators associated with each ecosystem, assessed according to the Common International Classification of Ecosystem Services [36]. The European methodologies followed for mapping ecosystems [36] use the interpretation of different land use and land cover classes and relate them to the European Habitat Classification (EUNIS). As a result, the analysis of ecosystem services while assessing land-use mapped ecosystems tends to present itself transformed into an "in situ" category, in the sense of [36], where the production of the service and the benefit are located in the same place. Another consequence of a methodology based on current land use mapping is to assess the use and service regardless of whether it is in an area of greater ecological suitability. Such is the example of forests that are all converted into ecosystems capable of producing the same services regardless of the type of forest species. This is an incorrect approach since it is well known that different forest trees or stands provide different ecosystem services. In Portugal, this is very relevant since most of the forest stands are not native (maritime pine, eucalyptus) providing poorer ecosystem services when compared with native stands (oak).

It is essential to consider that the landscape has different capabilities to provide specific ecosystem services [33], being of a more profound complexity than just an assessment by current land use. The structure of the ecosystem assumes a vital role in supporting the very functioning of the ecosystem. Interestingly, authors such as Burkhard [33] consider that the typology of supporting ecosystem services (defined in the Millennium Ecosystem Assessment), is understood as those that ensure ecological integrity. However, supporting services are considered difficult to map [37], and it is considered that the link between supporting services and human well-being occurs indirectly [34]. Therefore, supporting services have received much less attention among four types of ecosystem services. Despite European recommendations to map ecosystems and their services, recent publications have failed to include supporting services [38].

In this sense, it is considered that by mapping green infrastructure, the supportive services are provided in conjunction with other ecosystem services. Incorporating structural components of ecosystems also allows for a complete approach to mapping ecosystems and their services by encompassing the various relationships between the area of production of the service and the site of benefit from that service.

Including ecosystem services in landscape planning will need to go through defining the goal of achieving multiple ecosystem services [39]. However, the function of an ecosystem must be ensured in planning regardless of the benefit it may provide [40]. Maximizing one ecosystem service may jeopardize the balance of all other ecosystem services, as exemplified by Dosskey [39] in the US Green Belt region, where agricultural productivity was put as a priority at the expense of water quality and wildlife. It should be desirable for public policy to seek a degree of multifunctionality across cultural landscapes and to achieve the greatest degree of multifunctionality in green infrastructure [40]. The same author considers that monofunctional landscapes will require greater inputs to continue to provide values and functions and are likely to become unsustainable and require restoration.

Ecosystems are naturally multifunctional, making available services determined by landscape structure [39]. Modifying landscape structure can rebalance the services available. Thus, the landscape planning process will need to include an understanding of the current functioning of the landscape and an assessment of whether changing the landscape structure can affect the ecological functions and services provided [39]. Ecologically based planning aims to define the best use which implicitly includes the best use of natural resources without compromising their existence and their stability in the system [41]. This applies both to landscapes dominated by native vegetation, forming a dense woodland, and to agricultural or production forest areas properly integrated into the landscape and its structure with good management practices and appropriate design.

Ecosystem restoration is often focused on the recovery of a particular plant or animal community, often appearing related to landscape fragmentation [42]. The restoration of an ecosystem, which is itself a complex system, will involve the recovery of the ecological functions of the system. The functions that are easily altered by the degradation of an ecosystem are soil structure, nutrient flow, and water cycling [43]. The restoration of an ecosystem will mainly involve the recovery of the lost functions, because this loss has also contributed to its own degradation:

> *When a reintegrated landscape is achieved, it will be a landscape that is a mosaic of agricultural, natural, and semi natural systems, which together maximize biodiversity and economic returns by maintaining the landscape amenity function (minimizing the loss of landscape qualities through soil salinization of water and wind erosion) and so make possible a sustainable agriculture and a functional diverse natural system [43].*

The ecosystem restoration will involve the recovery of its ecological integrity. The integrity of an ecosystem includes the integrity of the system's structure and function, the maintenance of its components and its dynamic interactions [44]. From this perspective, any loss of system components leads to a loss of system integrity. According to Forman [45] and Thorn [46] a system with ecological integrity exhibits natural conditions of productivity, biodiversity, soil and water conservation, which are the goal of any sustainable environment [47]. The integrity of an ecosystem includes an adequacy of uses to the ecological characteristics of the system, meeting the dynamic stability of the landscape, making it resilient.

Ecosystem restoration is defined by the Society for Ecological Restoration as the process of supporting the recovery of an ecosystem that has been degraded, damaged, or destroyed [48]. These last three states can be equated to various states of morphogenesis, where an imbalance of landscape stability occurs. A morphogenesis ecosystem can be at different levels of imbalance from degradation to complete destruction. When ecosystems are being overexploited or degraded the health of the ecosystem goes into decline as well as its integrity and resilience [49]. This state of decline can be reversed with recovery actions, which in turn lead to ecosystem rebalance in Tricart's [3] interpretation of dynamic equilibrium. An ecosystem in equilibrium provides a greater number or a higher quality of services.

A Landscape-scale ecosystem restoration involves restoring a set of ecosystems to recover natural and cultural values and ecosystem service flows [49]. Hobbs [50] argues that the recovery of ecosystems should not focus on replicating the conditions prior to disturbance, but should be managed in a future perspective, not forgetting that the landscape is temporally and spatially dynamic. Besides, there are two types of recovery [51]: one in which the goal is the recovery of biotic continuity, and another, corresponding to more severe situations of degradation, will involve the physical recovery of the ecosystem. An example of the latter is the case of the obstruction of a river, where the goal will be to recreate the continuity of water flow. According to those authors, there are thus two thresholds that, if crossed, imply different interventions, the biotic threshold, and the abiotic threshold. The least severe situation of degradation of an ecosystem will thus imply changes in land use.

The ecosystem restoration can integrate cultural values, for example, an agricultural area located on productive soils and with good management practices (including compartmentalization hedges, contour farming) contributes to increased productivity, biodiversity, and soil and water conservation. Designing green infrastructure with planned land uses, consistent with its different ecological characteristics, guarantee the ecosystem's integrity and allow it to assess restoration needs.

3. The need for a restoration solution: centre region of Portugal

The absence of an ecological-based landscape and land-use plan can have severe consequences in the increase of soil degradation and floods, decrease of biodiversity, and increased fire risk. This landscape degradation is present in several areas of Portugal. Also, a consequence of the set of policies followed since the beginning of the 20th century. In the 1930s, Portugal went through a wheat campaign that destroyed the fertility of the soil. Later, the monoculture campaigns of maritime pine and eucalyptus continue the degradation of the soil and the destruction of the landscape, which we still see happening, especially in the north of the Tagus river, in the Centre and North regions.

The Centre Region (**Figure 1**) corresponds to a Statistical Territorial Unit (NUT) and comprises about 2,819,936 hectares. This region includes different landscape typologies, such as the southwest western zone with fruit productivity, the coastal zone with low and high coastlines, and an inland zone dominated by maritime pine and eucalyptus in formations of schist and granite. This area is characterized by a rugged relief in the most central location, such as Serra da Estrela, where Rio Mondego begins, the Serra da Lousã and the Serra do Açor. The Natura 2000 network includes 21 Special Areas of Conservation.

In an evaluation of land use from the 1990s to the present, it was possible to understand the evolution of the different land uses (**Figure 2** and **Table 1**).

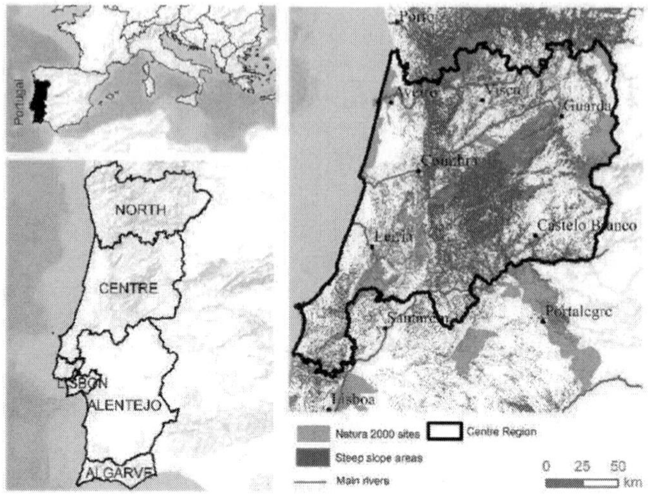

Figure 1.
Location of the Centre region in Portugal.

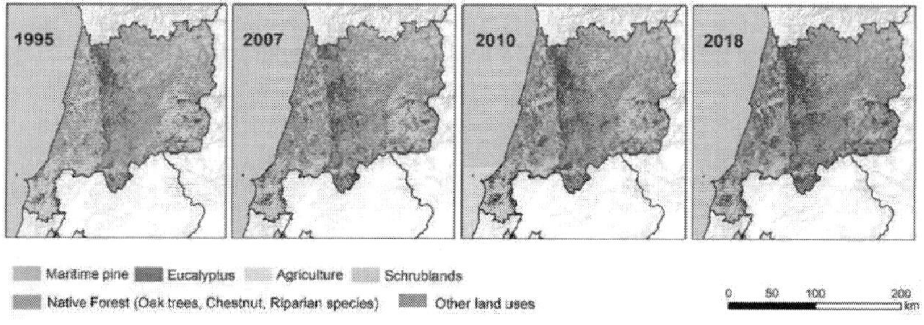

Figure 2.
Evolution of the main land use classes in the Centre region, between 1995 and 2018 (data base: DGT, land use and land cover map).

Main Land Uses in Centre Region	% in 1995	Variation	% in 2007	Variation	% in 2010	Variation	% in 2018
Maritime pine	26,7%	↘	20,0%	↗	22,5%	↘↔	22,3%
Eucalyptus	9,7%	↗	10,9%	↗	13,5%	↗	17,2%
Schrubland	11,7%	↗	16,4%	↘	12,3%	↗	13,3%
Agriculture	31,3%	↘	25,5%	↘	26,2%	↘	23,3%
Native	7,6%	↘	5,8%	↘	6,4%	↗	9,3%

Table 1.
Evolution of the main land uses in the Centre region, between 1995 and 2018.

This analysis was done with the interpretation and reclassification of the Land-Use and Land Cover maps produced by DGT (1995, 2007, 2010, 2018) [52]. In this period of 24 years, there is an oscillation of maritime pine, which tends to stabilize at 22%. The percentage of eucalyptus in the central zone has increased since 1995, from

9.7% of the total area, to 17% in 2018, i.e., it has practically doubled. On the other hand, the area occupied by agriculture has decreased since 1995. Native species include cork oak, holm oak, other oaks and also chestnut stands (archaeophyte), and oscillate between the years analyzed, with occupations between 6 and 9%.

There is in fact a very serious problem of inappropriate land uses that lead to the destruction of landscapes with negative consequences for those who live there, but also for those who live further away, for example, the impacts derived from water quality. Since the end of the 19th century, the oaks and chestnut trees and the traditional pastures, were replaced, first by maritime pine, which were planted on the community lands ("Baldios"), and then by eucalyptus. Land property fragmentation and, consequently, the landowners' increase also aggravated the land management problem.

A balanced landscape constituted by agriculture on the best soils, mixed woodland complementing agriculture and all its by-products, pastoralism in articulation with woodland and agriculture, villages, towns, and cities strategically located in situations of greater comfort and proximity to the food and materials produced was replaced by a landscape ecologically degraded, humanely depopulated and which burns extensively and repeatedly.

Fire frequency (number of times between 1990 and 2017)	Area (ha)	%
1 time burnt area between 1990 and 2017	698 775	24.8
2 times burnt area between 1990 and 2017	321 099	11.4
3 times burnt area between 1990 and 2017	93 965	3.3
4 times burnt area between 1990 and 2017	16 105	0.6
5 times burnt area between 1990 and 2017	2 882	0.1
6 times burnt area between 1990 and 2017	327	0.012
7 times burnt area between 1990 and 2017	224	0.008
No burned area between 1990 and 2017	1 686 559	59.8

Table 2.
Fire frequency between 1990 and 2017, area and percentage of case study.

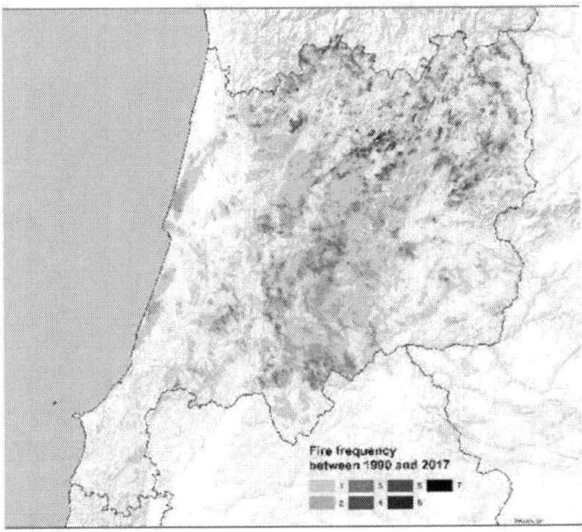

Figure 3.
Fire frequency between 1990 and 2017.

The policies disconnected from the ecological capacity of the land, led to the current situation of mega-fires, with loss of life and property and land abandonment. Analyzing the centre region in terms of burnet areas, 40% of the region was burned between 1995 and 2017. About 25% of the Centre Region burned once (**Table 2**), but 11% burned twice (**Figure 3**). In the megafires from 2017, the Centre Region was the most affected, representing 15% of total region area (416 thousand hectares). It is very urgent to develop adequate land-use plans for the rural areas.

The creation of a healthier landscape implies the conservation of natural resources, the creation of a balanced, multifunctional system with landscape recovery using native species. This will lead to creation of different economies, where ecosystem services payment can also take place.

4. The solution for healthier landscape – centre region of Portugal

The solutions developed to attain a healthier landscape involve the creation of a multifunctional landscape, with native or archaeophytes species, agriculture, and pastureland. The planned landscape will allow to create businesses, generate employment (landscape recovery companies, native species nurseries, forest management companies, reactivation of native wood business), and unique products (non-wood products, such as flour from oak acorns, chestnut, walnut, honey and mushroom production) capable of attracting nature tourism as well.

This landscape will then provide better ecosystem services, such as water quality, soil conservation, and biodiversity improvement, among others. Together, it will develop a fire resilient landscape, which is the landscape's capacity to absorb the disturbance caused by rural fires without losing its function, structure, and identity and ultimately weakening fire frequency and intensity or magnitude [53].

The vision for a healthier landscape of the Centre Region was developed by the application of FIRELAN [53], which is an ecologically based model that integrates different principles related to landscape fire resilience and ecological sustainability into a land-use plan, using the river basin as a landscape unit. The FIRELAN pretends to provide a multifunctional landscape (**Figure 4**) with benefits to the environment, but also developing economies and rural communities.

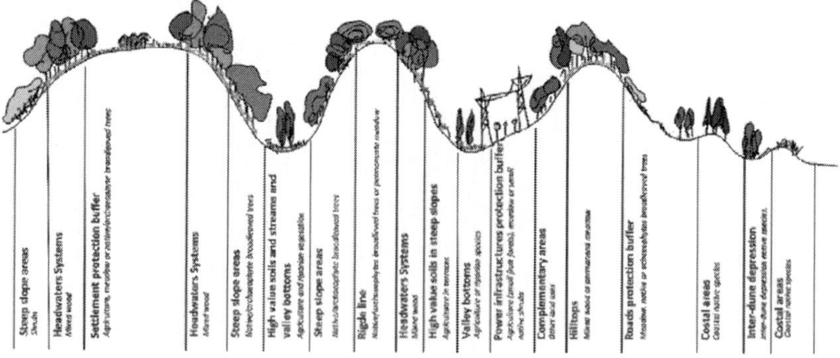

Figure 4.
Ecological and cultural system and land use potential from FIRELAN model.

The different components of the FIRELAN network for the Centre Region are mapped in **Figure 5**. The FIRELAN network is the main landscape structure with physical, biological, and cultural elements (**Table 3**). For each component there is a set of adequate land uses that should be promoted. Those land uses are identified in **Figure 6** and in **Table 4**. In the interstices of the FIRELAN network, also called Complementary Areas, the land use possibilities are wider.

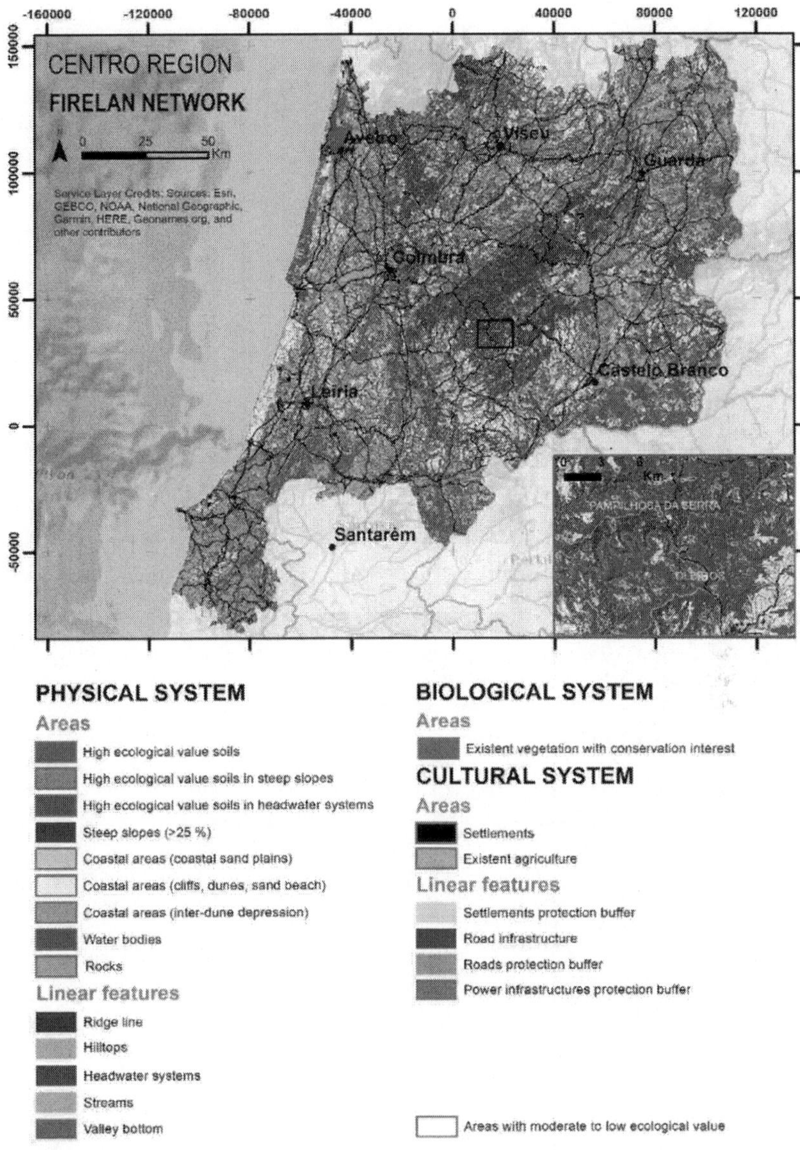

Figure 5.
FIRELAN network components in the Centre region of Portugal.

Firelan network component			Data Source
Physical System	Areas	Soils with high and very high ecological value	[54–56]
		Steep slopes (>25%)	[56, 57]
		Coastal area	[56, 58]
		Wetlands	[53]
		Water bodies	[53]
	Linear	Ridge line	[56, 57]
		Hilltops	[56, 57]
		Headwater systems	[56, 59]
		Streams and Valley bottoms	[56, 57]
Biological System	Areas	Existent vegetation with conservation interest	[52, 56, 60]
Cultural System	Areas	Settlements	[52]
		Existent agriculture	[52]
	Linear	Settlement protection buffer	—
		Roads	OpenStreetMap©
		Roads protection buffer	—
		Power infrastructures buffer	[61]

Table 3.
Components of FIRELAN and sources for the Centre region of Portugal.

The potential land uses plan (**Figure 6**, **Table 4**) highlights that:

- Native and archaeophytes species (including agroforestry systems), which represent 11% of the study area, can expand into further 31%.
- The agriculture can be increased in about 12% of the case study area, in addition to existing agriculture (23%).
- Existent vegetation with conservation interest is present in 14% of the Centre region.
- Complementary areas (areas with moderate to low ecological value) represent 16% of the study area.

Comparing potential land use and current land use map allows defining a Landscape Transformation Plan with conservation and restoration actions. According to the developed plan (**Figure** 7, **Table 5**): 35% of the Centre region should have restoration actions, and 57% should be maintained and conserved. Also, according to the results:

- The eucalyptus can be kept in 5% of the case study area, only in complementary areas, and with environmental measures.
- The maritime pine can be kept in 6% of the case study area, only in complementary areas, and with environmental measures.
- About 0,8% of the case study area should have recovery of the riparian vegetation.

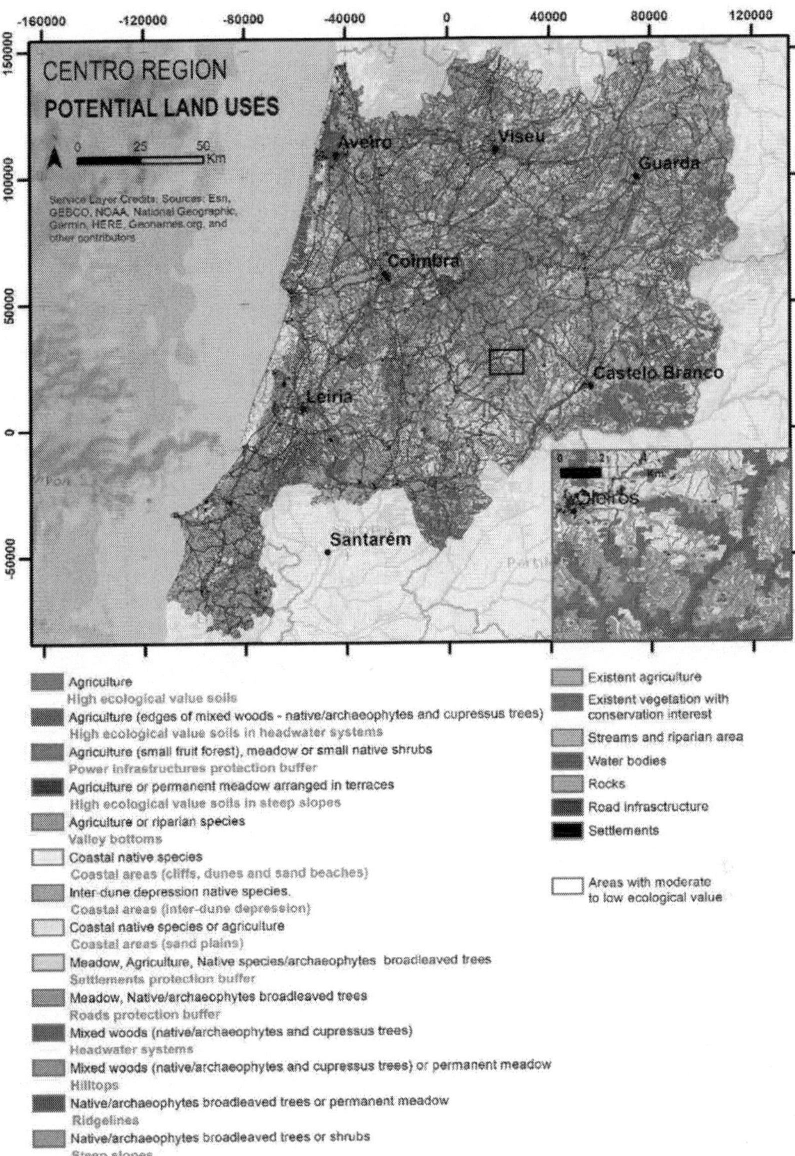

Figure 6.
Potential land uses in the Centre region of Portugal.

5. Conclusions

The United Nations General Assembly proclaimed the Ecosystem Restoration decade between 2021 to 2030, aiming to halt the degradation of ecosystems and restore them to achieve healthier landscapes. Landscape Architecture is an interdisciplinary discipline helpful in attaining those goals through planning and design

Firelan network components	Potential land uses	Area (ha)	%
High ecological value soils	Agriculture	131627	4.67
High ecological value soils in headwater systems	Agriculture (edges of mixed woods - native/ archaeophytes and cupressus trees)	52733	1.87
Power infrastructures protection buffer	Agriculture (small fruit forest), meadow or small native shrubs	5545	0.20
High ecological value soils in steep slopes	Agriculture or permanent meadow arranged in terraces	14408	0.51
Valley bottoms	Agriculture or riparian species	24395	0.87
Coastal areas (cliffs, dunes and sand beach)	Coastal native species	35653	1.26
Coastal areas (inter-dune depression)	Inter-dune depression native species	14178	0.50
Coastal areas (coastal sand plains)	Coastal native species or agriculture	5147	0.18
Settlements protection buffer	Meadow, Agriculture, Native species/ archaeophytes broadleaved trees	105883	3.76
Roads protection buffer	Meadow, Native/archaeophytes broadleaved trees	27454	0.97
Headwater systems	Mixed woods (native/archaeophytes and cupressus trees)	314469	11.15
Hilltops	Mixed woods (native/archaeophytes and cupressus trees) or permanent meadow	42221	1.50
Ridgelines	Native/archaeophytes broadleaved trees or permanent meadow	86392	3.06
Steep slopes	Native/archaeophytes broadleaved trees or shrubs	248223	8.80
Existent agriculture	Existent agriculture	646346	22.92
Existent vegetation with conservation interest	Existent vegetation with conservation interest	381180	13.52
Streams and riparian area	Streams and riparian area	22309	0.79
Water bodies	Water bodies	26709	0.95
Rocks	Rocks	3310	0.12
Road infrastructure	Road infrastructure	50764	1.80
Settlements	Settlements	132965	4.72
Areas with moderate to low ecological value	Areas with moderate to low ecological value	447731	15.88

Table 4.
Potential land uses for the Centre region of Portugal, area and percentage.

restoration in different contexts and scales, from rural to urban, from inland to coastal. The ecological-based planning methodologies contribute to better landscapes by starting from the knowledge and spatialization of natural processes. As part of an ecological-based planning methodology, working with green infrastructure allows the ecological integrity of the landscape, increasing the number or quality of services provided by the ecosystems and defining restoration actions and locations.

The Centre Region of Portugal has a severe problem of inappropriate land use. Since the end of the 19th century, the oaks and chestnut trees were replaced by

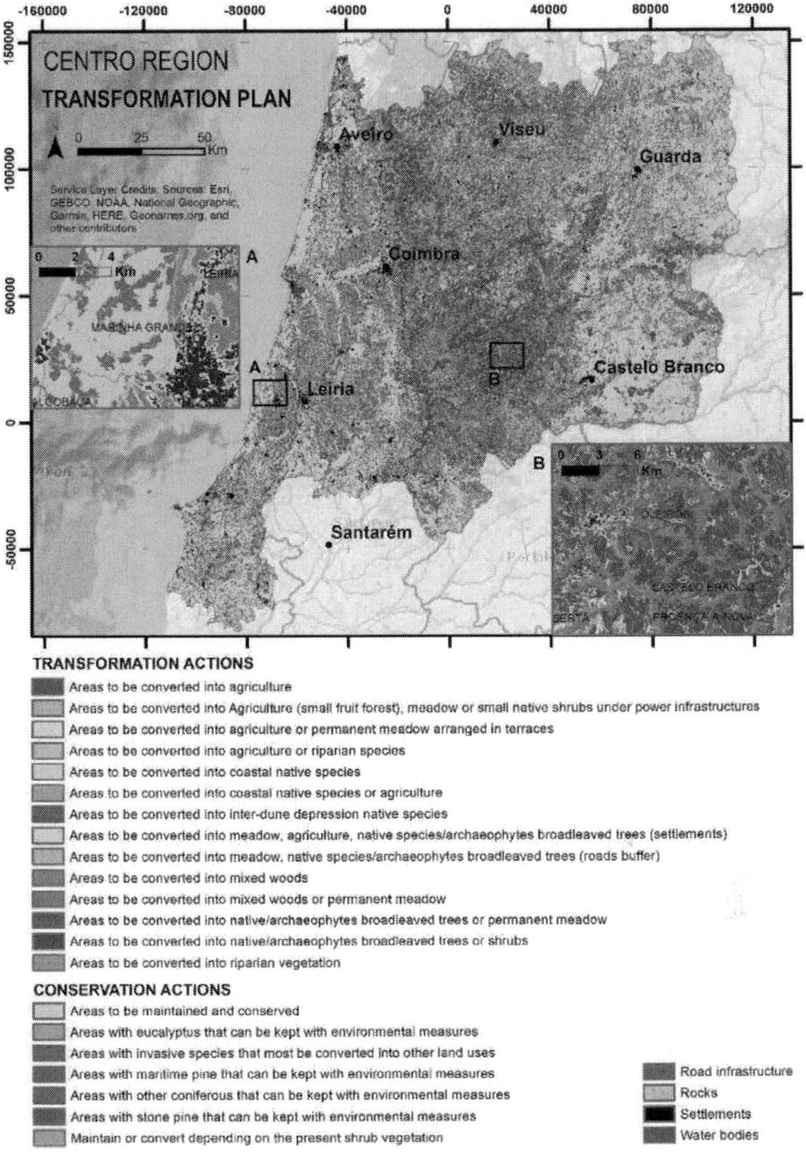

Figure 7.
Landscape transformation plan for the Centre region of Portugal.

maritime pine and eucalyptus. The number of fires and their severity increased, the land is continually degrading, and biodiversity loss increases. Applying an ecologically based landscape plan to prioritize restoration actions at the landscape scale is necessary to reverse this situation. The FIRELAN model was applied to the Centre Region to establish a vision for a healthier landscape. According to the results, 35% of the Centre Region should have restoration actions.

Landscape Transformation Actions		Area (ha)	%	Area (ha)	%
Restoration	Areas to be converted into agriculture	157670	5.59	980608	34.78
	Areas to be converted into Agriculture (small fruit forest), meadow or small native shrubs under power infrastructures	4332	0.15		
	Areas to be converted into agriculture or permanent meadow arranged in terraces	13938	0.49		
	Areas to be converted into agriculture or riparian species	18149	0.64		
	Areas to be converted into coastal native species	7840	0.28		
	Areas to be converted into coastal native species or agriculture	2786	0.10		
	Areas to be converted into inter-dune depression native species	14125	0.50		
	Areas to be converted into meadow, agriculture, native species/ archaeophytes broadleaved trees (settlements)	96844	3.43		
	Areas to be converted into meadow, native species/archaeophytes broadleaved trees (roads buffer)	24944	0.88		
	Areas to be converted into mixed woods	314469	11.15		
	Areas to be converted into mixed woods or permanent meadow	35461	1.26		
	Areas to be converted into native/ archaeophytes broadleaved trees or permanent meadow	80508	2.86		
	Areas to be converted into native/ archaeophytes broadleaved trees or shrubs	185648	6.58		
	Areas to be converted into riparian vegetation	22309	0.79		
	Areas with invasive species that most be converted into other land uses	1584	0.06		
Conservation	Areas to be maintained and conserved	1212952	43.02	1625284	57.64
	Areas with eucalyptus that can be kept with environmental measures	142909	5.07		
	Areas with maritime pine that can be kept with environmental measures	161113	5.71		
	Areas with other coniferous that can be kept with environmental measures	3283	0.12		
	Areas with stone pine that can be kept with environmental measures	2951	0.10		
	Maintain or convert depending on the present shrub vegetation	102077	3.62		
—	Road infrastructure	50764	1.80	—	—

Landscape Transformation Actions		Area (ha)	%	Area (ha)	%
—	Rocks	3310	0.12	—	—
—	Settlements	132965	4.72	—	—
—	Water bodies	26709	0.95	—	—

Table 5.
Transformation action, area and percentage of Centre region case study.

Subsequently, the eucalyptus area must drop from 17% of the Centre Region area to 5%, and the maritime pine from 22–6%. The results also show that agriculture could increase from 23–35% of the Centre Region. In the restoration actions, the native species should be used in more than 24% of the case study, especially in the headwaters systems and streams where mixed woods and riparian galleries should be developed.

The landscape transformation plan contributes to the definition of adequate policies to tackle ecosystem restoration through landscape and land use planning.

References

[1] Cabral FC. O "Continuum Naturale" e a Conservação da Natureza. (The "Continuum Naturale" and nature conservation.) Proceedings of the seminar Conservação da Natureza, Lisboa: Serviços de Estudos do Ambiente, 18 e 19 de Abril de 1980. p. 35-54

[2] Telles GR. 1992. A conservação das Paisagens Históricas e Rurais. Correio da Natureza 17. 1992 p. 53-55

[3] Tricart J. Ecodinâmica. Rio de Janeiro, IBGE, Diretoria Técnica, SUPREN. 1997. Available from: https://www.docdroid.net/D8HPCLd/jean-tricart-ecodinamica-pdf [Accessed: 2016-05-02]

[4] Tricart J. Morphogènese et pédogenèse, géomorphologie et pédologie. Science du Sol 1, 1965. P. 69-85

[5] Odum EP. Fundamentals of ecology. Saunders College Publishing, Philadelphia, PA (US), 1971. 574 pp.

[6] ONU. Convention on biological diversity. 1992. Available from: https://www.cbd.int/doc/legal/cbd-en.pdf [Accessed: 2015-09-01]

[7] COP 5. Fifth Ordinary Meeting of the Conference of the Parties to the Convention on Biological Diversity, 15 - 26 May 2000 - Nairobi, Kenya. Available from: https://www.cbd.int/decisions/cop/?m=cop-05 [Accessed: 2015-09-01]

[8] Magalhães MR, Abreu MM, Lousã M, Cortez N. Estrutura Ecológica da Paisagem. Conceitos e delimitação – escalas regional e municipal. ISAPress, Lisboa, 2007. 361pp.

[9] Magalhães MR, Cunha N, Pena S. The Ecological Land Suitability in the Land-Use Plan. In Proceedings of the 7th International Conference on Virtual Cities and Territories, Nova University, TULisbon, October 11-13, 2011, Portugal. 2011. p. 307-312

[10] Collins MG, Steiner FR, Rushman MJ. Land-use suitability analysis in the United States: historical development and promising technological achievements. Environ. Manage. 2001; 28 (5), 611-621. DOI: 10.1007/s002670010247

[11] Magalhães MR,. A Arquitectura Paisagista. Morfologia e Complexidade. Editorial Estampa, Lisboa, 2001. p.525.

[12] McHarg IL. Design with Nature. NHP, New York (US), 1971. 197 pp.

[13] Steiner FR. The Living Landscape, Second Edition: An Ecological Approach to Landscape Planning, Island Press. 2008.

[14] Firehock K. Strategic Green Infrastructure Planning. A multi-scale approach. Island Press. 2015. 138pp.

[15] Kolbe J, WUstemann H. Estimating the Value of Urban Green Space: A hedonic Pricing Analysis of the Housing Market in Cologne, Germany. J. Chem. Inf. Model. 2014. 5, 45-61. DOI:10.1017/CBO9781107415324.004

[16] Costanza R, d'Arge R, De Groot R, Farber S, Grasso M, Hannon B, Limburg K, Naeem S, O'Neill RV, Paruelo J, Raskin RG, Sutton P, Van Den Belt M, The value of the world's ecosystem services and natural capital. Nature 1997. 387, 253-260. DOI:10.1038/387253a0

[17] Hermann A, Schleifer S, Wrbka T. The Concept of Ecosystem Services Regarding Landscape Research: A Review. Living Rev. Landsc. Res. 2011. 5, 1-37. DOI:10.1177/0170840609104565

[18] Bennett G, Mulongoy KJ. Review of experience with ecological networks, corridors and buffer zones. 2006. Secretariat of the Convention on Biological Diversity, Montreal, Technical Series No. 23.

[19] Taylor PD, Fahrig L, Henein K, Merriam G. Connectivity is a vital element of landscape structure. Oikos. 1993. 68, 571-573. DOI:10.2307/3544927

[20] MacArthur Rh, Wilson EO. The theory of island biogeography. Princeton Univ. Press. Ed., Princeton. 1967.

[21] Cunha NS, Magalhães MR. Methodology for mapping the national ecological network to mainland Portugal: A planning tool towards a green infrastructure. Ecological Indicators 2019. 104C: 802-818. DOI: 10.1016/j.ecolind.2019.04.050

[22] EEA. Green infrastructure and territorial cohesion. 2011. Technical report No 18/2011

[23] Bennett AF, Haslem A, Cheal DC, Clarke MF, Jones RN, Koehn JD, Lake PS, Lumsden LF, Lunt ID, Mackey BG, Nally RM, Menkhorst PW, New TR, Newell GR, O'Hara T, Quinn GP, Radford JQ, Robinson D, Watson JEM, Yen AL, Ecological processes: A key element in strategies for nature conservation. Ecological Management & Restoration, 2009. 10: 192-199. DOI: 10.1111/j.1442-8903.2009.00489.x

[24] Forman RTT, Godron M. Landscape ecology. John Wiley, New York. 1986. 619p.

[25] Naumann S, McKenna D, Timo K, Mav P, Matt R. Design, implementation and cost elements of Green Infrastructure projects. Final report to the European Commission, DG Environment, 2011. Contract no. 070307/2010/577182/ETU/F.1, Ecologic institute and GHK Consulting.

[26] EC. Our life insurance, our natural capital: an EU biodiversity strategy to 2020. COM(2011) 244 final. 2011.

[27] EC. Towards an EU Research and Innovation policy agenda for Nature-Based Solutions & Re-Naturing Cities. 2015. DOI:10.2777/765301

[28] EC. EU Biodiversity Strategy for 2030 Bringing nature back into our lives COM/2020/380 final. 2020.

[29] EC. Green infrastructure (GI): enhancing Europe's natural capital COM(2013) 249 final. 2013.

[30] De Groot R, Brander L, van der Ploeg S, Costanza R, Bernard F., Braat L, Christie M, Crossman N, Ghermandi A, Hein L, Hussain S, Kumar P, McVittie A, Portela R, Rodriguez LC, ten Brink P, van Beukering P. Global estimates of the value of ecosystems and their services in monetary units. Ecosyst. Serv. 2012. 1, 50-61. DOI:10.1016/j.ecoser.2012.07.005

[31] Boyd J, Banzhaf S. What are ecosystem services? The need for standardized environmental accounting units. Ecol. Econ. 2007. 63, 616-626. DOI:10.1016/j.ecolecon.2007.01.002

[32] Müller F, Burkhard B. The indicator side of ecosystem services. Ecosyst. Serv. 2012. 1, 26-30. DOI:10.1016/j.ecoser.2012.06.001

[33] Burkhard B, Kroll F, Müller F. Landscapes' Capacities to Provide Ecosystem Services – a Concept for Land-Cover Based Assessments. Landsc. Online 2010, 1-22. doi:10.3097/LO.200915

[34] Alcamo J. Ecosystems and human well-being, Ecosystems. ISLAND PRESS, Washington, D.C., USA. 2003. p. 245.

[35] Fisher B, Turner RK, Morling P. Defining and classifying ecosystem

services for decision making. Ecol. Econ. 2009. 68, 643-653. doi:10.1016/j.ecolecon.2008.09.014

[36] EU. Mapping and assessment of ecosystems and their services in the EU. Indicators for ecosystem assessments under Action 5 of the EU Biodiversity Strategy to 2020, 2nd Report. 2014. DOI:10.2779/75203

[37] Benis E, Drakou E, Dunbar MB, Maes J, Willeman L. Indicators for mapping ecosystem services: a review. 2012. DOI:10.2788/41823

[38] EEA. European ecosystem assessment— concept, data, and implementation Contribution to Target 2 Action 5 Mapping and Assessment of Ecosystems and their Services (MAES) of the EU Biodiversity Strategy to 2020. 2015. DOI:10.2800/629258.1.

[39] Dosskey MG, Wells G, Bentrup DW. Enhancing ecosystem services: Designing for multifunctionality. J. Soil Water Conserv. 2012. 67, 37A – 41A. DOI:10.2489/jswc.67.2.37A

[40] Selman P. Planning for landscape multifunctionality. Sustain. Sci. Pract. Policy, 2009. 5, 45-52. DOI: 10.1080/15487733.2009.11908035

[41] Ndubisi F. Managing change in the landscape: a synthesis of approaches for ecological planning. Landscape Journal. 2002. University of Wisconsin Press. EBSCO Publishing. p. 138-155

[42] Jaeger JAE, Soukup T, Madriñán LF, Schwick Chr, Kienast F. Landscape fragmentation in Europe. EEA-FOEN Report. 2011. ISBN 978-92-9213-215-6

[43] Main AR. Landscape Reintegration: Problem Definition. In: Richard J. Hobbs, Denis A. Saunders (editors) Reintegrating Fragmented Landscapes. 1993. Springer New York. p 189-208.

[44] Regier HA,. The Notion of Natural and Cultural Integrity. In: Woodley, S., Key, J., Francis, G. (editors) Ecological Integrity and the Management of Ecosystems. USA. St. Lucie Press. 1993. p3-18

[45] Forman RTT. Land Mosaics. Cambridge University Press, Cambridge, 1995. p. 632.

[46] Thorne JF. Landscape ecology. A Foundation for greenway design. In Paul S.Smith & Paul Cawood Hellmund (editors) "Ecology of greenways" University of Minnesota press. 1993. p 23-42

[47] Cook EA. Landscape structure indices for assessing urban ecological networks. Landsc. Urban Plan. 2002. 58, 269-280. DOI: 10.1016/S0169-2046(01)00226-2

[48] SER, 2004. The SER International Primer on Ecological Restoration, in: Www.ser.org & Tucson: Society for Ecological Restoration International.

[49] Clewell AF, Aronson J. Ecological Restoration. Principles, Values, and Structure of an Emerging Profession. 2ª edition. USA. ISLANDPress. 2013

[50] Hobbs RJ, Hallett LM, Ehrlich PR, Mooney H a,. Intervention Ecology: Applying Ecological Science in the Twenty-first Century. Bioscience, 2011. 61, 442-450. DOI:10.1525/bio.2011.61.6.6

[51] Hobbs RJ, Harris JA. Restoration Ecology : Repairing the Earth ' s Ecosystems in the New Millennium. Restor. Ecol. 2001. 9, 239-246. DOI:10.1046/j.1526-100x.2001.009002239.x

[52] DGT, Land Use and Land Cover Maps from different years: 1995, 2007, 2010, 2018. Available from: https://snig.dgterritorio.gov.pt/ [Accessed: 2020-09-01]

[53] Magalhães MR, Cunha NS, Pena SB, Müller A. FIRELAN—An Ecologically

Based Planning Model towards a Fire Resilient and Sustainable Landscape. A Case Study in Center Region of Portugal. Sustainability 2021. 13, 7055. DOI:10.3390/su13137055

[54] Leitão M, Cortez N, Pena SB. Solo (Soil). In Magalhães, M.R.M. (editor) Estrutura Ecológica Nacional. Uma Proposta de Delimitação e Regulamentação (National Ecological Network. A Delimitation and Regulation Proposal). ISAPress, Lisboa. 2013. p. 83-104, ISBN 978-972-8669-53-9.

[55] Pena SB, Abreu MM, Magalhães MR, Cortez N. Water erosion aspects of land degradation neutrality to landscape planning tools at national scale. Geoderma, 2020. 363, 114093. DOI: 10.1016/j.geoderma.2019.114093.

[56] EPICWebGIS (http://epic-webgis-portugal.isa.ulisboa.pt/)

[57] Cunha NS, Magalhães MR, Domingos T, Abreu MM, Withing K. The land morphology concept and mapping method and its application to mainland Portugal. Geoderma, 2018. 325: 72-89. DOI: 10.1016/j.geoderma. 2018.03.018.

[58] Silva JF, Cunha N, Lopes AM, Abreu MM, Magalhães MR. Litoral (Coastal area). In: Magalhães, MRM (editor), Estrutura Ecológica Nacional. Uma Proposta de Delimitação e Regulamentação (National Ecological Network. A Delimitation and Regulation Proposal), ISAPress, Lisboa. 2013. p. 67-82ISBN 978-972-8669-53-9.

[59] Pena SB, Magalhães MR., Abreu MM. Mapping headwater systems using a HS-GIS model. An application to landscape structure and land use planning in Portugal. Land Use Policy 2018. 71, 543-553. DOI: 10.1016/j. landusepol.2017.11.009

[60] Mesquita, S. (2013). Vegetação (Vegetation). In: Magalhães MRM (editor), Estrutura Ecológica Nacional. Uma Proposta de Delimitação e Regulamentação (National Ecological Network. A Delimitation and Regulation Proposal), ISAPress, Lisboa. 2013. p. 105-120. ISBN 978-972-8669-53-9

[61] REN. Rede nacional de transporte de electricidade (National electricity transport network), 2012, Portugal. Escala 1:1.000.000. REN - Rede Elétrica Nacional, S.A. Lisboa

Chapter 6

The American Landscape Architecture Research Universe and a Higher Education Ordination: Descriptive Insights into the Discipline and Profession of Landscape Architecture

Abstract

Landscape scholars, educators, and academics are interested in the structure and nature of the knowledgebase that comprises both the discipline of landscape architecture and the profession of landscape architecture. In this study, the latent nature of the landscape architecture discipline was revealed by constructing a principal component citation analysis representation (the landscape architecture research universe) concerning several decades of literature (1982–2017) in *Landscape Journal*, a preeminent American journal addressing landscape architecture research. In addition, an ordination was developed describing the curriculum relationships between fifteen top American universities teaching landscape architecture as identified by 'DesignIntelligence,' preparing students for practicing in the profession of landscape architecture. The results revealed that in the discipline, the research activity is highly diverse along many dimensions, constantly evolving as new topics are explored. The pattern in landscape architecture research is broad, as the discipline integrates knowledge and ideas in many fields. In contrast, landscape architecture curriculums, teaching the fundamentals of the profession, are fairly closely clustered together and quite similar, with small differences reflecting emphasis in either landscape history or the visual arts, and mathematics or course electives. This dual identity is both a source of conflict and a unique opportunity.

Keywords: information science, multivariate analysis, environmental design, higher education, citation analysis

1. Introduction

Over the past few decades, information scientists, plus others, have been interested in the structure and nature of the knowledgebases that comprises both disciplines/professionals in many academic areas and subjects, including landscape architecture. A discipline is the body of information collected, studied, analyzed, and reported by a group of individuals who collectively are affiliated with a subject [1]. For the most part, a discipline is usually associated with being a science—describing the way of the universe as best as it can be deciphered, interpreted, and explained, usually with the scientific method. On the other hand, a profession is an activity where a group of individuals practice the art of the profession—making decisions about what to do and how to accomplish the task. For example, in the area of medicine, researchers study the body, conduct experiments, and report results in the discipline of medicine. In contrast, medical doctors give advice to patients and perform operations, deciding what to do and what should be done, often without perfect information, practicing the art of medicine. Usually those who study the discipline are found at research institutions and organizations. Those who practice the profession are typically in business applying their art. Doctors, lawyers, architects, planners, musicians, and athletes are all examples of practitioners applying their skill, deciding what to do and what should be done; thus, it is called the 'art of practice.'

In landscape architecture, dominance has been expressed through the activities of the profession, where individuals practice the art of decision making for planning, design, construction, and maintenance of the exterior environment [1]. It was only relatively recently (1980s) that any attention was given to the discipline of landscape architecture, although some may claim landscape research extends back at least to the a thesis by Frank Waugh concerning campus planning and design at Oklahoma State University for a master's degree at Kansas State University in 1894 [2–6]. The debate concerning the difference between professional practice and the need for the accreditation of schools offering professional practice degrees and the role of research in graduate education is illustrated in *Graduate Education in Landscape Architecture: a Compendium* [7]. Much has changed since the 1980s in the discipline of landscape architecture. An undated report by the American Society of Landscape Architects (ASLA) illustrates how little activity in landscape research was undertaken by American landscape architecture programs in the United States in the early 1970s [8]. A report titled: *Metrics Evaluating Multivariate Design Alternatives: Application of the Friedman's Two-way Analysis of Variance by Ranks: A Personal Reflection*, provides some insight into the progression and develop of landscape research over the last 50 years from the viewpoint of one American scholar [9].

Research in landscape architecture can be divided into two aspects. The first is the development of predictions (models) [1]. Models can be equations, graphs, or even 3-dimensional representations. The other aspect is the development of theories (explanations) [1]. In addition, theories can be further divided into scientific theories (explanations about the universe that if shown to be false are discarded) and normative theories (explanations about reasons and ideas forming a a foundation for decision making, such as a set of ideas about why a designer created a design in a particular manner—exceptions can always be found and all of these normative theories are false, but they are not scientific theories and are simply guides or principles to make decisions in an imperfect world of knowledge—for a designer this is very useful) [1]. There is very little in the way of scientific theory in landscape architecture as most of the theory is normative, useful for practitioners. Most books on landscape theory are about normative theory, ideas and approaches for creating and managing landscape. For example, the deployment of a concept in a design is a normative theory [10, 11]. In contrast, landscape scholars often focus their energy upon developing predictive models accepting the models as evidence but rarely focusing upon scientific theory. Examples of predictive models developed by landscape architects are in human perception research related to assessing visual quality [12–15] and in natural resources to develop soil reclamation Eqs. [16–21]. 'Human intrusion theory' in explaining visual quality equations [22] and 'mesic preference theory' for reclaiming surface mines [23] are explanations that are scientific theories developed by landscape scholars. Within this context/framework of models and theories, landscape research has evolved.

As the volume of landscape research accumulated. Research about research was of interest to some. One approach to study this research was to derive a structure to examine citations in articles written and published in journals [24]. This general approach was reported in a study by Dr. Burley and his spouse Cheryl (an information scientist) concerning the landscape architecture literature for a journal titled: *Landscape Journal* [25]. The co-authors of this book chapter queried Dr. Burley about the reception of this effort. "Well, for the conference, they gave us a premiere setting at the beginning of the conference. With the exception of a few conference people in the room who were required to attend the session, the room was empty. There were a lot more people in the hotel bar. At the time, I really do not think anyone went to these conferences to learn about research, but rather to escape their academic institution, converse with friends and colleagues, and unwind. No one was interested. A better venue would have been an information science setting. Still, I kept looking for opportunities to expand the research endeavor. I was undaunted, very independent; I still am." reflected and commented Dr. Burley. In 2009, a similar expanded study was reported examining the landscape research literature in transportation [26]. Surprisingly, this study was noticed and featured in a seminal book about landscape research [27]. And the study earned an ASLA state award for research. An interesting finding in the study was that the results indicated a fractured, weakly linked research universe where investigators were deep into their line of research and not tied or integrated into other areas. In contrast, in the landscape architecture discipline, there were many connections and interrelationships. "The blending and borrowing across different subject areas was something that landscape architects have claimed for a long time. The study supported those claims. Often in academia, other disciplines tout their depth and wonder why landscape architects do not do the same? Again, here was evidence that in one area, environmental transportation, they were deep but unconnected. I believe both approaches are beneficial, but the differences illustrate where conflicts from those who believe in one approach over the other can be generated. Because landscape architects borrow and integrate, it can go unappreciated by other academics." assessed Dr. Burley.

The foundation of the research is to employ multivariate principal component analysis (PCA), something that landscape architects rarely study. "During my time as a graduate student, my professors at the University of Manitoba urged me to take as many advanced statistic courses as possible and I took even more at the University of Michigan for my PhD. It was like learning the analytic tools for conducting research. If one does not know the tools, it can be difficult to understand the possibilities. Similarly in landscape architecture, if one does not know the design process, it is difficult to generate a design. Many landscape programs around the world have research programs, but seem to emphasize learning more about the environment and less about the tools of research." noted Dr. Burley. In ecology and other fields, multivariate analysis was essential to study and compare settings and ideas. Curtis studied vegetation communities in Wisconsin and ordinated the communities by recording the frequency, density, and dominance of each plant type in a stand [28]. An ordination of research activity can be accomplished by treating the category of literature cited in an article (like a vegetation type) and the article itself as a stand of vegetation. "When it was first proposed to me about studying research structure of literature with citations, it only took me about 15 seconds to develop the experimental design, but it had taken half a lifetime to be prepared for those 15 seconds." stated Dr. Burley. With this basic analytic tool (PCA), other kinds of studies related to garden design, cemeteries, cultural heritage landscapes, and paintings have been examined by those working with Dr. Burley [29–34].

The intent of the study reported in this book chapter, an expanded investigation of the landscape research literature to visualize the changes across time for *Landscape Journal* were initiated. The study provides insight into how topics studied change and evolved.

2. Methodology

For this investigation, the analysis reveals a latent underlying structure for the landscape architecture discipline (the landscape architecture research universe) concerning the citation literature of *Landscape Journal* from several decades of articles (1982–2018). *Landscape Journal*, is a preeminent American journal addressing landscape architecture research and is affiliated with the Council of Educators in Landscape Architecture (CELA).

For each issue, the study team collected all of the peer reviewed published articles, 'source articles' for the study years. Each source article comprised one observation set. For one observation set there would usually be journal articles cited in the bibliography. These cited journal articles contained within the bibliography of a source article are called 'citation articles'. To classify a citation article, the Library of Congress classification number for the journal title of each citation article was recorded. If the same journal is cited more than once, the tally will be greater than one. Within an observation set, the total number of citation articles for a particular category was tallied. For example, if the subject category 'architecture' had 6 cited architecture citation articles in a source article, the architecture tally for the observation set would be six. The Library of Congress classification was chosen as it was an existing, broad, and easy to use system, recognized by many major state research universities. The Library of Congress system is non-hierarchical, meaning that the new bodies of knowledge that emerge are relatively easily incorporated into the classification system and thus as the system grows over time, it can accommodate modifications and development in the knowledge base. Flexibility over time was an essential component since historical research may span across a wide time frame.

In this study there were 38 subject variables. Thus each observation sets had 38 scores, each representing the tally of each subject from the source article. With the subject areas for all of the journals identified, one could then sort the citation articles from each source article into a subject category. Citations to literature such as monographs, technical reports, and books were not included in this study. I addition, proceedings were included only if they appeared to be published at least annually, meaning it was a serial. Once the subject areas for each source article were tabulated, the dataset could then be entered into a computer for statistical analysis.

Multivariate data analysis was performed using SAS 9.1 [35]. To conduct a PCA, the subject categories were each standardized to a mean of 0.0, standard deviation of 1.0. The standardization is important to the analysis [36]. Otherwise, the results will be dominated by categories with large scores. After standardization principal component analysis can be conducted upon the observation sets (an observation set is comprised of the scores in 38 subject category variables for a source article). The analysis produces a numerical table present eigenvalues which represent independent dimensions, from the largest value to the smallest. For interpretation, eigenvalues for standardized data with values over 1.0 were considered significant dimensions. The significant dimensions represent bodies of knowledge in the landscape research university. Significant dimensions were selected for further analysis by examining the eigenvector coefficients of each dimension which indicate the level of association that a subject category had with the dimension. In other words, eigenvector coefficients numerically illustrate the correlation between a variable (the subject category) and the dimension. The eigenvector coefficients are arranged in a table, sorted by the eigenvalue and would range in score between −1.0 and 1.0. Values near 0 indicate low correlation with the eigenvalue dimension while values near −1.0 or 1.0 indicate a strong association with the dimension. In this study, eigenvector coefficients with a value greater than or equal to 0.400 or less than −0.400 were considered to be affiliated with a particular dimension. Subject categories with more than one significant eigenvector coefficient meant that the subject was significant across more than one dimension, suggesting a dimension connecting subject category. Subject categories with only one significant eigenvector were considered primary to the associated eigenvalue. Primary categories were employed to label (name of identify) a dimension. Weak associations with the dimensions were considered to be eigenvector coefficients ranging from −0.4 to −0.20 and 0.20 to 0.4. The results of the PCA were plotted creating a structural map (universe) of the dimensions, associated subject categories, and connecting subjects. In other words, this map could graphically describe the latent properties of the data. The map would be a graphical depiction of the research universe in a given time frame. Several time frames were examined: the complete time frame from 1982 to 2018, 1997 to 2007, 1999 to 2009, 2001 to 2011, 2003 to 2013, 2005 50 2015, and 2007 to 2017.

In contrast to the research universe, an ordination was also developed describing the curriculum relationships between fifteen top American universities teaching landscape architecture as identified by 'DesignIntelligence,' preparing students for practicing in the profession of landscape architecture [37]. Each school was an observation set and the subjects taught in the curriculum were the categories in each observation set. The categories were standardized, and PCA invoked. The results of the latent dimension can be plotted to illustrate the relative position of one school to another. The intent is not to show which is better, but rather to identify similarities and differences. The plots can depict an educational univers in a manner similar to other types of plots [29–34].

3. Results

The results for the complete set of source articles studied (1982–2017) indicated that at least 16 dimensions were significant, meaning they had eigenvalue scores greater than 1.0 (**Table 1**). The eigenvector coefficients for the first four dimensions are included in **Table 2** to illustrate eignevectors from the tables. The complete tables are too extensive to print in this book chapter; however, they are available from the corresponding author. Across different time frames, the number of dimensions ranged for 14 to 17significant dimensions. The large number of dimensions suggest a fair number of topics are being studied within the profession. There is great diversity in what landscape scholars study and what comprises the breadth of the landscape discipline.

Twenty-two subjects were found in the study of the curriculums for the fifteen top 2016 undergraduate school in the United States, PCA analysis revealed that the subjects could be compacted into fourteen dimensions (**Table 3**). **Table 4** illustrates the first two eigenvector coefficients for the first two eigenvalues from **Table 3**.

Dimension	Eigenvalue	Dimension	Eigenvalue
1	3.37416400	20	0.86118014
2	2.37773394	21	0.82850031
3	2.30449507	22	0.79618238
4	1.72251904	23	0.76786013
5	1.57488980	24	0.73251054
6	1.48665358	25	0.70882083
7	1.35165038	26	0.70309916
8	1.28649532	27	0.60824656
9	1.24473535	28	0.60049263
10	1.19781171	29	0.58875709
11	1.14476595	30	0.56176491
12	1.11569426	31	0.53288311
13	1.08994852	32	0.49631484
14	1.05809264	33	0.46083954
15	1.03103935	34	0.42602992
16	1.01171969	35	0.38829465
17	0.98221308	36	0.36183629
18	0.96307548	37	0.32252224
19	0.93616755	38	0.00000000

Table 1.
Eigenvalue scores for the set of source articles from 1982 to 2017.

Categories	Prin1	Prin2	Prin3	Prin4
General works, newspapers, college publications	–.007198	–.002037	*–.310990*	*0.367060*
Psychology/ environmental psychology/ esthetics/ethics	0.042329	*0.302969*	0.021038	–.070908
History	–.075144	–.129145	*0.307017*	*0.367516*
Geography	0.031008	0.149171	0.119874	0.142929
Human ecology, anthropogeography	0.081835	0.098209	–.011167	0.057008
Anthropology/ folklore	–.036767	–.055084	0.141117	0.176610
Recreation/ leisure	–.043616	0.257909	0.133548	–.074602
Social sciences, general	–.056506	*0.333947*	0.004893	0.074859
Economics	0.038984	*0.375711*	0.101212	0.127883
Economics/business	–.028212	0.095794	0.085080	–.043298
Sociology	0.014627	**0.439660**	0.065118	0.061568
Law	0.011909	0.215752	–.020283	0.174770
Education	–.028176	–.037113	0.231629	0.006814
Visual arts	–.031139	–.024711	–.210592	–.046823
Architecture	–.044855	0.106333	–.183968	0.013958
Planning:	–.050149	0.270738	–.000570	–.082755

Categories	Prin1	Prin2	Prin3	Prin4
Science	*0.399670*	−.059300	0.082008	−.044951
Mathematics/computer science	−.028045	−.036028	0.003420	−.138301
Physics/meteorology	0.087573	−.057693	−.042900	−.106514
Geology	0.190450	−.077314	*0.321455*	0.002699
Natural history/ecology	**0.410378**	0.045850	0.015143	−.006438
General biology, zoology, botany	*0.389709*	0.001750	0.142236	0.031558
Agriculture	*0.340821*	0.044523	0.024557	0.049860
Plant sciences	−.017463	−.054205	−.132844	−.088730
Landscape architecture	−.050347	0.225171	−.044795	−.172863
Forestry	0.095814	0.209337	−.070056	0.203556
Technology/engineering	0.068859	0.061201	−.222159	0.284676
Library science	−.014933	−.029205	0.024091	−.037774
Language and literature	−.058329	−.019355	−.122869	−.030357
Military	−.023128	−.080632	−.163174	*0.352283*
Community health, medicine, nursing	*0.376630*	−.028026	−.250960	−.012459
Political sciences	*0.376630*	−.028026	−.250960	−.012459
Africa ecology	0.139597	−.071225	*0.363935*	0.017311
Religion, theology	0.032129	0.151920	0.117238	0.139279
Archeology, genealogy, civilization	−.042456	−.146986	0.195003	**0.412185**
Decorative arts	−.027927	0.047571	−.166330	0.110450
Physiology	−.010245	0.132418	0.042379	−.047434
Environmental sciences	−.026603	−.059235	−.116184	0.265003

*Note: *Bold* coefficients in red indicate categories with a *strong* association for a particular principal component (dimension); *Italic* blude coefficients indicate categories with *a modest* association for a particular principal component (dimension); *Underlined* coefficients indicate categories associated with more than one dimension.*

Table 2.
Eigenvector coefficients for the first four dimensions from source articles 1982 to 2017.

4. Discussion

4.1 Landscape research universe

The plotting and description concerning all the time frames examined would be longer than allowed for the space allotted to this book chapter. However three time periods from and the universe of research the educational program universe are of particular interest (**Figures 1–4**). For the decade from 1992 to 2002, the research universe had expanded to many more dimensions from 10 in as first reported by Burley and Burley to 16 with environmental science as the largest dimensions giving way to agriculture [25]. Yet by the decade from 1998 to 2008, agriculture gave way to a more amorphous environmental science dimension and a total of 17 dimensions within the universe. The trend for amorphous categories continued until the dominant dimension in 2006–2016 was an amorphous unlabeled dimension. This suggests that some of landscape research was clustered in undefined and uncategorized set, defying description. For some this may be refreshing and or others this may be disturbing. While the categories change and the size of them varies, the complexity remains

Eigenvalues of the Covariance Matrix				
	Eigenvalue	**Difference**	**Proportion**	**Cumulative**
1	4.488223	1.336905	0.1951	0.1951
2	3.151318	0.154661	0.137	0.3322
3	2.996657	0.213096	0.1303	0.4624
4	2.783561	0.550254	0.121	0.5835
5	2.233306	0.506218	0.0971	0.6806
6	1.727088	0.097973	0.0751	0.7557
7	1.629115	0.473776	0.0708	0.8265
8	1.155339	0.28917	0.0502	0.8767
9	0.86617	0.29367	0.0377	0.9144
10	0.5725	0.072126	0.0249	0.9393
11	0.500374	0.029942	0.0218	0.961
12	0.470432	0.233565	0.0205	0.9815
13	0.236867	0.047816	0.0103	0.9918
14	0.189051	0.189051	0.0082	1

Table 3.
Eigenvalue scores for the set of subjects studied at the 15 schools.

	Prin1	**Prin2**
Psychology	0.321304	0.194173
History	0.403998	0.070797
Geography	−0.02829	−0.00676
Anthropology	0.003392	−0.38156
Social sciences	0.123925	0.021618
Economics	−0.15176	0.295427
Sociology	−0.02066	−0.12461
Visual arts	−0.23055	0.162509
Architecture	0.011708	0.154676
Planning	0.271037	0.243036
Mathematics/computer science	−0.09343	0.446772
Physics	0.399808	0.055758
Geology	−0.04178	−0.00577
Natural history/ecology	0.39242	−0.00639
General biology, zoology, botany	−0.24822	0.074506
Agriculture	−0.0826	−0.1556
Plant sciences	−0.02562	−0.06083
Landscape architecture	−0.24405	0.145394
Language and literature	−0.08476	−0.04105
Political Sciences	0.266506	0.152063

	Prin1	Prin2
Electives (dark matter)	0.046689	−0.51369
Humanities	−0.16304	0.149757
Chemistry	−0.08505	0.174573

*Note: *Bold* coefficients in red indicate categories with a *strong* association for a particular principal component (dimension);*

Table 4.
Eigenvalue scores for the subjects in the first wo dimensions. Studied at the 15 schools.

across the time frames. In any one time frame, much of the remaining research not placed in a dimension, representing the proportion of research not placed in an significant dimension is 35.84 percent (the sum of eignevalues in **Table 1** that are less than one and then divided by 38) of the research activity. This means that about 1/3rd of the research activity is not in a cluster and not categorized. There is a fair amount independent exploration.

Are **Figures 1–3**, what one expects to see? or desires to see? Some may call for a more unified focus and other may call for even more anarchy and diffusion in landscape research.

4.2 Landscape education universe

In comparison to **Figures 1, 2,** and **3, Figure 4** presents a very different universe. Landscape architecture dominates with 54.9 percent of the subject material taught and in second place it the amorphous dark matter of electives which defy categorization. **Table 5** illustrates the average percent of academic categories taught at the five schools.

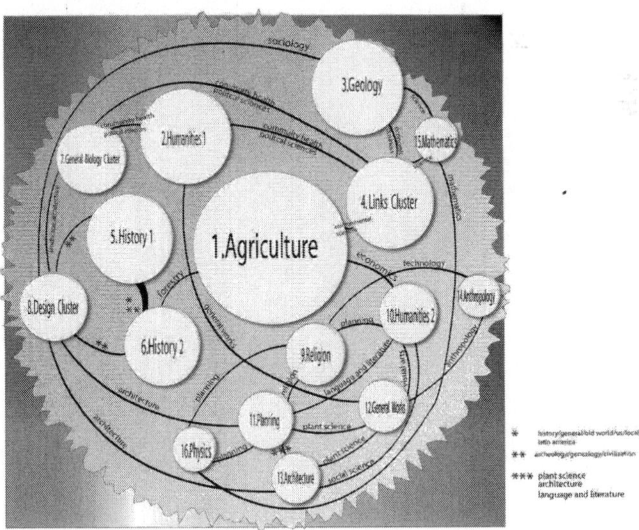

Figure 1.
A drawing of the landscape research universe from 1992 to 2002.

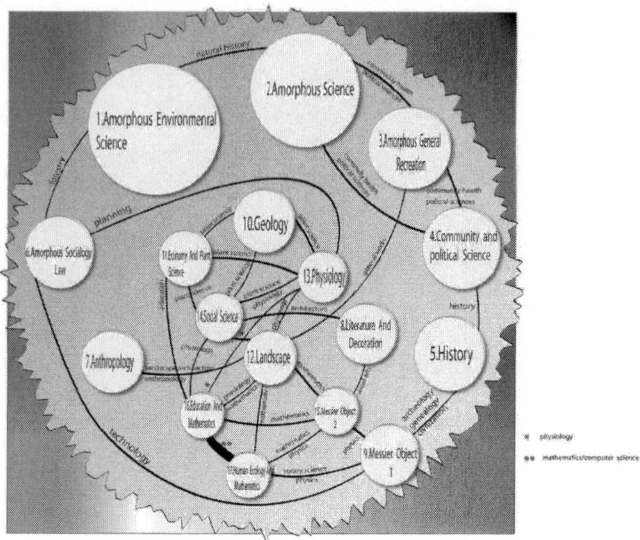

Figure 2.
A drawing of the landscape research universe from 1998 to 2008.

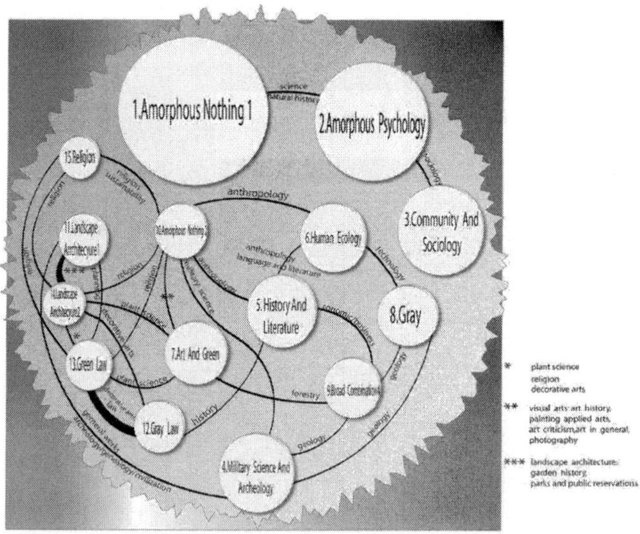

Figure 3.
A drawing of the landscape research universe from 2006 to 2016.

Figure 5 is an ordination plot of the fifteen schools based upon the first two dimensions. All fifteen schools are very good schools and share much in common. However, some schools emphasize one area over another. While the MSU landscape architectural program is not highly mathematical or visual in emphasis, it is relatively more than its peer institutions. If the schools were drastically different, the scale on the dimensions would be in the tens not the single digits.

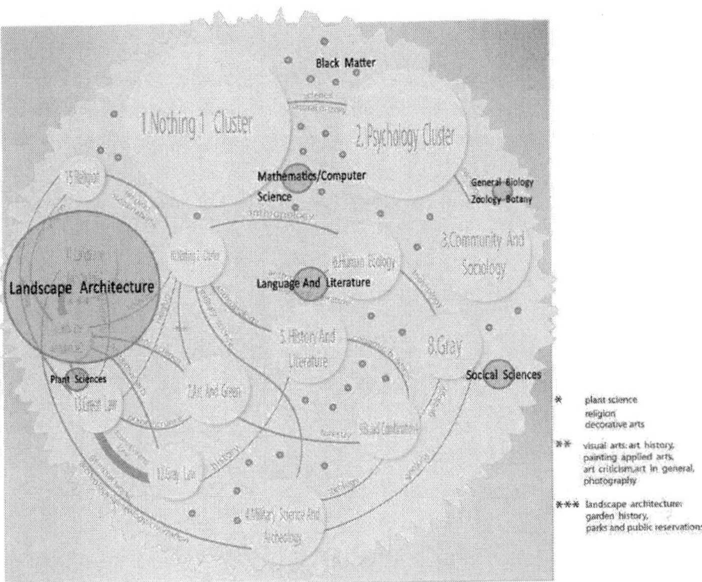

Figure 4.
A drawing of the landscape architectural program education universe 2016 overlaid upon ***Figure 3.***

Category	Average
Psychology	0.16%
History	1.2%
Geography	0.26%
Anthropology	0.41%
Social sciences	2.42%
Economics	0.57%
Sociology	0.91%
Visual arts	1.67%
Architecture	1.28%
Planning	1.25%
Mathematics/computer science	3.81%
Physics	0.92%
Geology	0.75%
Natural history/ecology	0.79%
General biology, zoology, botany	2.33%
Agriculture	0.47%
Plant sciences	2.59%
Landscape architecture	54.91%
Language and literature	5.66%
Political Sciences	0.56%

Category	Average
Electives	15.04%
Humanities	1.78%
Chemistry	0.26%

Table 5.
Percentage of subject categories taught at the top 15 American undergraduate schools.

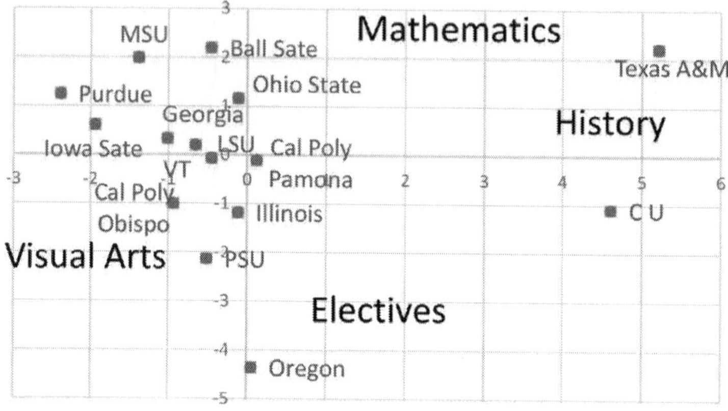

Figure 5.
An ordination plot of the fifteen school based upon the first two dimensions (dimension one horizontal, dimension 2 vertical).

4.3 Blending the two universes

It becomes apparent, that to teach landscape architecture and to do research in landscape architecture occupy two different realms. This understanding is not new to those who work in the academic treadmill. However, the results revealed in this study supports this belief. A tenure stream academic in landscape architecture may have to balance two worlds: the more narrow focus of training landscape architects and the extremely broad and diverse world of landscape research. And it is not surprising that many landscape architecture undergraduates would have little connection or interest in research. And it is not surprising that many newly hired professors coming from their professional training would be unprepared to tackle a research endeavor.

The co-authors asked Dr. Burley about his observations concerning this change, as he has observed, witnessed, and participated in this transition. "I believe much of the change began in the 1970s. The push for research has to do with money and university ranking. Schools around the world are now competing with each other for status and position. Administrators compete for a ranking, as the ranking is based upon publications, citations, and money. Therefore, administrators need to coerce/urge their faculty to obtain grants, publish, and be cited. I was told by someone who had been a faculty member in the MSU department of geography that in the mid-1970s, it was very rare for anyone to have a grant, although in their department many published. In the 1970s very few published, if ever in landscape architecture at MSU. This was frustrating for MSU administration. The merits of landscape architecture in service to society are admirable; however, these merits do not contribute to university ranking and comparatively, teaching landscape architecture as a major is expensive (small classrooms, dedicated studio space). The landscape program at Michigan State University was slated for closure in the 1980s because the faculty were slow to adopt the mission of publication and grant writing. Then Dr. Jo Westphal was hired in the landscape program and the transition began. The hiring of myself and Dr. Mary Ann Kniseley was the second phase of that transition. To explain further, university priorities change based upon where the money is to be found. In the past the state often funded many public schools, but that money has long disappeared at many institutions. The money has been found by raising tuition quickly and by seeking eternal funding for research. Majors in the humanities and general education may give way to majors in medicine and physics. Schools change their identity. It is in this academic environment landscape architecture educators and students find themselves placed within. It is easy to imagine where conflict can arise and also where opportunities may exist. In the 1970s, the mindset of landscape architecture academics was purely in a setting similar to **Figure 4**. But now on the research side of things it has evolved in the last 40 years to something similar to **Figure 3**. I was a part of that change. I am not saying it is for the good and the better, nor suggesting that it is negative either. That is for others to decide. I am too close to the middle of it to make a judgment. But indeed, it has been fun to discover and uncover measures and analytic approaches to understand what has happened and to work with fine colleagues from around the world." stated Dr. Burley.

The co-authors also asked Dr. Burley about his interpretation of what this change means for landscape architecture faculty. "Well, first it is a source of conflict at many levels. I have witnessed it many times around the world at many landscape architecture programs and in discussions with many faculty. To illustrate how successful this has been, not one hired landscape architecture assistant professor has made it to full professor at MSU in over the last 40 years. That is a tragic track record. But it really does not matter from the university's perspective because very few know this track record--administrators and most faculty have a very short time frame in their positions. The two deans who were recently hired to oversee our department/school did not even finish one full term—they left. When I was hired, there were four of us as new assistant professors in a multi-disciplinary department, but after 8 years, I was the only one who remained, the rest had left. Of the last nine landscape faculty to leave the landscape program over the last 30 years, I can say all left somewhat disgruntled, jaded and often disillusioned. I am sure it will be no different for me. Yet the university can present a positive perspective to the outside world. From the thousands of professors it hires, it only needs to show possibly 20 or so success stories each year to market the university in a very positive manner (that is 4/10ths of one percent of the instructor population at MSU). In the 30 years I have been at MSU, rarely are individuals in my department/school ever featured. It has happened; however planning and design scholars are not a priority (remember in the 1980s they tried to dispose of this group) and not what the university may wish to project as an image. Often, I see publications featuring laboratories and medicine. There is often an optimistic attitude about the future. While past events may have resulted in dismal failure, the belief that the next person hired will bring a bright and happy future is a consistent theme. Then reality sets in, problems occur, people leave, and the bright and happy future of the next forthcoming hire is all that is discussed. Over 40 years ago at another institution, I would listen to a certain dean's yearly report to stakeholders. He would always paint a bright and beautiful future. But after several years of this, I would recall the new initiatives he had promoted the years before, most ending in an unpleasant manner. But it did not matter, no one remembered them (but I did). All that mattered was that the forthcoming year was going to be marvelous. Universities struggle with this all the time. The quest for money, publications, and citations at an ever increasing level generate many internal problems. And because most universities accept this incremental race for recognition and prestige, in many ways they generate the problems and issues that arise at the institution. As has been said before, 'We have met the enemy and it is us.' While universities may claim to be bastions upholding diversity and equity; often instead, they are halls of elitism, intolerance, insecurity, and arrogance with no chance for true inclusion and diversity – in fact it can be quite brutal. This is often what I have observed for landscape architecture faculty at many institutions around the world. Still many try, and some do succeed; but one will rarely hear about the many who did not succeed. I am not attempting to present a dreary image, but rather I have been in academic for over 45 years and at one institution for nearly 30 years, plus have lectured at around 35 universities world-wide, and at many more conferences, so eventually one gets an understanding of what is occurring. **Figures 3** and **4**, make a lot of sense to me. They help to explain the setting and the situation." noted Dr. Burley.

To cope with this duality, one approach that universities have been employing is something known by some as the 'Stanford Academic Educational Model.' The model establishes two classes of instructors and researchers. In the Stanford Academic Educational Model, one academic class of employees, the researchers, are highly paid, in the tenure stream, teaching only advanced graduate student courses, focus upon producing research papers in the most highly respected journals possible, usually seeking research grants to support their efforts. The other academic class of employees, the instructors, are paid at a lower level, not in the tenure stream, teaching the masses of undergraduate students, have no research responsibilities, and are not required to produce journal articles. The researchers may have a very high opinion of themselves and the instructors will wonder why the researchers are not as engaged with the students. It is the difference between **Figures 3** and **4**. At some institutions and within departments this causes great internal strife and battles. The differences are reflected in the expectations of those serving educational professional practice and those serving the search for new knowledge. Universities attempt to be entrepreneurial with their research faculty and still serve the needs of the student body, searching for relevance, contributions, and meaning for the public [38–40].

"At MSU, it used to be that most of the faculty were a blend between the two types, one in the tenure stream, conducting research, writing papers, and teaching all levels of students. There were very few employees in the purely instructional model. But MSU has drifted towards the Stanford Academic Educational Model where now about half of the faculty are instructors. This approach saves the university substantial salary money. Since the instructors are not in the tenure stream, it brings administrators more flexibility to hire, fire, and change academic offerings/majors. It used to be that the instructors were not even considered faculty, but with about half of the academics now being non-tenured stream, universities have found means to label them as faculty. Titles are easy to give/anoint and cost almost nothing. And it would not help the university's cause for it to be known that the number of what had been known as faculty staffing had been reduced in half. Universities struggle to find approaches that still serve their student body clientele bringing in tuition dollars and striving to maintain their academic ranking and position with journal articles, citations, and research dollars. I find neither fault nor praise for what has transpired, but rather based upon the differences between **Figures 3** and **4**, I understand why this has happened. At one time there was an interesting documentary film shown on American Public Television, describing the struggles and challenges of one part-time instructor at Stanford, but I have been unable to find a citation for this film. It is very revealing. Stanford has a very well respected Department of Art and Art History which produces many excellent documentaries." reflected Dr. Burley.

"The push to maintain university rank and standing continues. In the past most educators in landscape architecture had master's degrees. I recall one European nation that urged landscape faculty to each earn a PhD., using termination as a stick and the promise of increased pay as a carrot. Eventually most earned a PhD. Upon conclusion, there no terminations but also was no money to support pay increases, but the faculty had earned PhD.s and started writing and publishing as part of their duties—mission accomplished. Faculty need to recognize that the goals and requirements for faculty by universities are going to constantly change, universities are going to expect more, not less and it will be driven by the need to sustain ranking and status above all else. For landscape architecture faculty, they need to understand that their existence is based both on the expectations of planning and design professionals to produce students illustrated in **Figure 4**, and to conduct research illustrated in **Figure 3**. And they need to understand that they are in competition with other departments and professional schools in their university. According to recent metrics in GoogleScholar, the top landscape architectural citated author in the world, William Sullivan at the University of Illinois, he has over 13,000 citations as of July 2021; but for example as my institution, the top cited authority was Joey Hudson a physicist, with over 336,000 citations. It was not until one reached down to about the 140th cited researcher that one was at the 13,000 metric. Approaching 1,000 citations, I am usually in the top 40 of cited landscape architecture researchers in the world, but I am not even close to the top 500 at my university. Universities look at these standings. It is not easy for landscape architecture faculty to compete in such an environment with the other departments and professional schools. When universities make decisions about where to invest, it is easy to understand their priorities. **Figures 3** and **4** offer insight into those challenges." observed Dr. Burley.

5. Conclusion

This study revealed that the landscape research universe has become complex with many dimensions growing and diminishing but always remaining complex. In contrast the landscape educational universe is more simplified and not congruent with the organization of the research universe. For landscape architecture academics, many reside in both of these dual worlds. Interest in this topic continues with recent published articles by Ozdil and by Newman (et al.) [41, 42].

References

[1] Burley, J.B., Machemer, T. From Eye to Heart: Exterior Spaces Explored and Explained. 1st ed. San Diego, California, USA: Cognella Academic Publishing; 2016.

[2] Zube, E.H. Research and design: prospects for the 1980s. In: Alanen, A. R. editor. Conference Proceedings: Research in Landscape Architecture: Selected Papers Presented at the Annual Meeting of The Council of Educators in Landscape Architecture, August 20-23, 1980. Madison, Wisconsin: Department of Landscape Architecture, School of Natural Resources, College of Agriculture and Life Sciences & The council of Educators in Landscape Architecture, 1981. p. 1-11.

[3] Miller, S.C. Threshold to perspective: Developing research in landscape architecture. In: Alanen, A. R. editor. Conference Proceedings: Research in Landscape Architecture: Selected Papers Presented at the Annual Meeting of The Council of Educators in Landscape Architecture, August 20-23, 1980. Madison, Wisconsin: Department of Landscape Architecture, School of Natural Resources, College of Agriculture and Life Sciences & The council of Educators in Landscape Architecture, 1981. p. 12-21.

[4] Chenoweth, R.E. Research in Landscape architecture: an extended definition. In: Alanen, A. R. editor. Conference Proceedings: Research in Landscape Architecture: Selected Papers Presented at the Annual Meeting of The Council of Educators in Landscape Architecture, August 20-23, 1980. Madison, Wisconsin: Department of Landscape Architecture, School of Natural Resources, College of Agriculture and Life Sciences & The council of Educators in Landscape Architecture, 1981. p. 22-30.

[5] Zube, E.H. The roots of future innovations: research and theory. Landscape Architecture; 1980. p. 614-617.

[6] Burley, J.B. Charles Parker Halligan's impact on the MSU Landscape Architecture program. LandTEXTURE. E. Lansing, Michigan: Michigan State University Extension; 2020.

[7] Vanlandingham, M.C., editor. Graduate Education in Landscape Architecture: a Compendium. Washington, D.C.: American Society of Landscape Architects Council on Education; 1987.

[8] American Society of Landscape Architects, Council on Education. Research in Schools of Landscape Architecture. Washington, D.C.: American Society of Landscape Architects, Council on Education; undated.

[9] Burley, J.B., Li, X., He, S. Metrics Evaluating Multivariate Design Alternatives: Application of the Friedman's Two-way Analysis of Variance by Ranks: A Personal Reflection. Perrinton, Michigan: Whitemud Academics; 2020.

[10] Loures, L., Burley, J. B. Conceptual Precedent: Seven Historic Sites Revisited. WSEAS Transactions on Environment and Development. 2009; 5(1):55-64.

[11] Burley, J.B. The design concept: intellectual landscapes in Michigan. The Michigan Landscape. 2009; 49(12):33-40.

[12] Jin, Y., Burley, J.B., Machemer, P., Crawford, P., Xu, H., Wu, Z., Loures, L. 2018. The Corbusier dream and Frank Lloyd Wright vision: cliff detritus vs. urban savanna. In: Ergen, M. editor. Urban Agglomeration. Rijeka, Croatia: Intech; 2018. p. 211-230.

[13] Burley, J. B., Yilmaz, R. Visual quality preference: the Smyser index variables. International Journal of Energy and Environment. 2014;8:147-153.

[14] Mo, F., Le Cléach, G., Sales, M., Deyoung, G., Burley, J. B. Visual and environmental quality perception and preference in the People's Republic of China, France, and Portugal. International Journal of Energy and Environment. 2011;4(5): 549-556.

[15] Burley, J.B. Visual and ecological environmental quality model for transportation planning and design. Transportation Research Record. 1997;1549:54-60.

[16] Yue, Z., Burley, J.B.. Predictive Models for Reforestation and Agricultural Reclamation: A Clearfield County, Pennsylvania Case Study. Vegetation Index and Dynamics. London, United Kingdom: Intech Open; 2021. p. 1-16.

[17] Wen, B., Burley, J.B. Soil-based vegetation productivity model for Coryell County, Texas. Sustainability. 2020;12(5240):1-14.

[18] Burley, J.B., Wu, Z., He, S., Li, X. Soil-based vegetation productivity models for disturbed lands along the northern and central, western Great Plains, USA. Journal of Advanced Agricultural Technologies. 2020;7(1):1-7

[19] Corr, D.L, Burley, J. B., Cregg, B., Schutzki. R. 2020. Soil based vegetation productivity models for reclaiming northern Michigan landscapes: a monograph. Journal of the American Society for Mining and Reclamation. 2020;9(4):1-43. DOI: http://dx.doi.org/10.21000/JASMR20040001

[20] Bai, Y., Chang, Q., Guo, C., Burley, J. B., Partin, S. Neo-sol productivity models for disturbed lands in Wisconsin and Georgia, USA. International Journal of Energy and Environment. 2016;10:52-60.

[21] Burley, J.B., Thomsen, C., Kenkel, N. Development of an agricultural productivity model to reclaim surface mines in Clay County, Minnesota. Environmental Management. 1989;13(5):631-638.

[22] Burley, J.B. The science of design: green vegetation and flowering plants do make a difference: quantifying visual quality. The Michigan Landscape. 2006;49(8):27-30.

[23] Burley, J.B., editor. Environmental Design for Reclaiming Surface Mines. Lewiston, New York: Edwin Mellen Press; 2001.

[24] Small, H., Griffith, B. C. The structure of scientific literature. I: identifying and graphing specialties. Science Studies. 1974;4(1):17-41.

[25] Burley, J. B, Burley, C. J. (1998) Citation analysis of landscape architecture research literature: the generation of a discipline's structural domain typology. In: Conference Abstracts, The University of Texas at Arlington, Council of Educators in Landscape Architecture. Arlington, Texas: University of Texas Arlington; 1998. p. 51-55.

[26] Burley, J.B., Singhal, V.B.P., Burley, C.J., Fasser, D., Churchward, C., Hellekson, D,. Raharizafy, I. Citation analysis of transportation research literature: a multi-dimensional map of the roadside universe. Landscape Research. 2009;34(4):481-495.

[27] Deming, M.E., Swaffield, S. Landscape Architectural Research: Inquiry, Strategy, Design. Hoboken, New Jersey: Wiley; 2010.

[28] Curtis, J.T. Vegetation of Wisconsin: an Ordination of Plant Communities. Madison, Wisconsin: University of Wisconsin Press; 1959.

[29] Chen, D., Burley, J. B., Machemer, R. Schutzki, R. Ordination of selected traditional Japanese gardens, traditional Chinese gardens, and modern Chinese gardens. International Journal of Culture and History. 2021;8(1):14-51. DOI: https://doi.org/10.5296/ijch.v8i1.18250

[30] Li, N., Wang, L., Jin, X., Yue, Z., Machemer, T., Zhou, J., Burley, J. B. An ordination of selected artists, painters, and designers: line, composition, color. Journal of Architecture and Construction. 2020;3(1):37-51.

[31] Loures, L., Burley, J. B., Panagopoulos, T., Zhou, J. Dimensions in post-industrial land transformation on planning and design: a Portuguese case study concerning public perception. Journal American Society of Mining and Reclamation. 2020;9(3):14-43. DOI: http://dx.doi.org/10.21000/JASMR20030014

[32] Wen, B., Burley, J. B. Expert Opinion Dimensions of Rural Landscape Quality in Xiangxi, Hunan, P.R. of China: Principal Component Analysis and Factor Analysis. Sustainability. 2020;12(1316).

[33] Xu, H., Burley,J.B., Crawford, P., Schutzki, R. Cross-cultural ordination of burial sites. International Journal of Cultural Heritage. 2017;2:92-104.

[34] Xu, Y., Burley, J. B., Machemer, P., Allen, A. A dimensional comparison between classical Chinese gardens and modern Chinese gardens. WSEAS Transactions on Environment and Development. 2016;12(2016):200-213.

[35] SAS. SAS 9.1 Cary, NC: SAS Institute Inc.; 2003.

[36] Johnson, R. A., Wichern, D. W. Applied Multivariate Statistical Analysis, 2nd edition. Englewood Cliffs, New Jersey: Prentice Hall; 1988.

[37] Cramer, J. P., editor. Landscape Architecture: Top 15 Programs. Design Intelligence: America's Best Architecture and Design Schools. 2026;21(6):43.

[38] Etzkowitz, H. StartX and the 'Paradox of Success': Filling the gap in Stanford's entrepreneurial culture. Social Science Information. 2013;52(4),605-627.

[39] Etzkowitz, H. Innovation in innovation: The triple helix of university-industry-government relations. Social science information. 2003:42(3),293-337.

[40] Etzkowitz, H. Entrepreneurial scientists and entrepreneurial universities in American academic science. Minerva. 1983;21(2-3),198-233.

[41] Ozdil, T. R. Who will teach the next generation of landscape architects? Ten-year review of academic position descriptions in landscape architecture in North America. Landscape Journal. 2020;39(1):55-69.

[42] Newman, G., Li, D., Tao, Z., Zhu, R. Recent trends in LA-based research: a topic analysis of CELA abstract content. Landscape Journal. 2020;39(2):51-73.

Index